Sampling Inspection Tables
Single and Double Sampling

Sampling Inspection Tables
Single and Double Sampling

SECOND EDITION
Revised and Expanded

HAROLD F. DODGE
and
HARRY G. ROMIG

Wiley Classics Library Edition Published 1998

A Wiley-Interscience Publication
JOHN WILEY & SONS, INC.
New York   Chichester   Weinheim   Brisbane   Singapore   Toronto

Copyright © 1944, 1959 by Bell Telephone Laboratories, Incorporated. All rights reserved.
Copyright © 1998 by Lucent Technologies Bell Laboratories.

Published by John Wiley & Sons, Inc.

Wiley Classics Library Edition published 1998.

Published simultaneously in Canada.

Library of Congress Cataloging-in-Publication Data:

Library of Congress Catalog Card Number 59-6763
0471-25549-1 (Classics Edition)

Printed in the United States of America.

10 9 8 7 6 5 4 3 2 1

Preface for the Wiley Classics Library Edition

Few books have contributed as much to the past, present, and future of statistical quality control as the Dodge–Romig, "Sampling Inspection Tables." Under the guidance and supervision of George D. Edwards, a small group of pioneers at Bell Telephone Laboratories, including Walter A. Shewhart, Harold F. Dodge, and Harry G. Romig, working together provided the basis of modern statistical process control and acceptance sampling. Well-used copies of the tables can be found on the shelves of a host of contemporary practitioners. They can be expected to be used extensively for years to come.

It is important to note that the original tables were developed at Western Electric at the time of Shewhart's investigation of the control chart. Their use requires estimation of the process level, which, of course, would be expected to be accomplished with a control chart. Contemporary emphasis on process control makes the tables particularly useful, since they provide a means for disposition of product when the process is out of control. Processes *do* go out of control, that is why we keep control charts. The question is what to do with the product in an out of control situation. These tables adjust the amount of inspection appropriately. Use with process control is part of their heritage. In fact, an excellent discussion of use of the tables with process control will be found on page 32, which concludes:

> . . . the use of these sampling procedures and tables together with continuing control chart analysis of the inspection results obtained therefrom, have been found to provide a balanced and economical inspection program.

This edition was written in 1959. It is just as true today. Incompetence, ignorance, and fraud make it necessary for a producer to check output of a process as well as the process itself. In so doing, the process will be given the attention it deserves, while product is protected from unexpected and undetected deterioration of quality.

The text accompanying the tables provides insight into statistical consideration involved in the construction of these plans as well as their operation. The tables have been checked many times by computer and have repeatedly been verified. They are just as relevant today as they were when they were developed. In any but the most basic industries, processes can be found with levels of fraction defective which meet or exceed the values given in the tables. Use of these plans while introducing process control can lead to process improvement and product conformance.

Mark Twain once said that a classic is a book which people praise and don't read. The Dodge–Romig Tables provide an outstanding example of the converse of this remark. For they have been read and reread by quality professionals for more than 50 years and are classic in every sense. They now have found a home as a Wiley Classics Library Edition, a tribute to the book, its authors and John Wiley & Sons itself, who had the foresight to make these tables available to an audience outside of Western Electric with the first edition in 1944. The tables have met the test of time and provide a sound basis for sampling for the future.

EDWARD G. SCHILLING, PH.D

December 1997

Preface to the Second Edition

THE publication of this second edition permits the inclusion of certain additional features that experience has shown to be desirable. Some of the new items are these:

(1) The Introduction has been expanded by the addition of two sections. One of these discusses the application of the tables when rejected lots are not screened, and the other outlines some of the steps that the individual worker must take in choosing a sampling plan for a specific application.

(2) A new chapter, Chapter 4, discusses the subject of the "operating characteristic curve" (OC curve) of a sampling plan.

(3) Appendix 1, new, presents OC curves for all the sampling plans in the single sampling AOQL tables.

(4) Appendix 2, also new, gives OC curves for all the sampling plans in the double sampling AOQL tables.

(5) In addition, Appendix 3 provides OC curves for a general set of single sampling plans for sample sizes $n = 500$ or less and acceptance numbers $c = 0$, 1, 2, and 3, based on binomial probabilities.

The content of the original text in Chapters 1, 2, and 3 remains substantially unchanged. Minor alterations have been made in the notation to eliminate a few inconsistencies among chapters; the words "defect" and "defective" have been clearly set apart, which was not consistently so in the original writings; and a few footnotes of comment and suggestion have been added where appropriate in lieu of modifying the text.

The addition of the OC curves in Appendices 1, 2, and 3 has been urged over the years by a number of engineers who have found the AOQL concept and designation to be the most useful of the various methods of indexing sampling plans.

Effective presentation of the OC curves necessitated a new size and format for the book. This enabled us to work with Mr. Kenneth M. Collins as he planned ways and means for turning out a product with emphasis on good design. The present typographic arrangement of the text matter and the tables was designed by Mr. Collins, who also suggested the system now used for numbering the sections, figures, and equations. We are indebted to him for his help in these and other matters connected with the preparation of this edition.

Much of the credit for the new material in this edition goes to associates in Bell Telephone Laboratories. We gratefully acknowledge Miss Mary N. Torrey's contributions to the technical content and her assistance in all phases of preparing the new edition, including the meticulous reading of proofs. Acknowledgments for the extensive and painstaking work of preparing the OC curves are given at the end of Chapter 4. Special mention goes to Mr. M. K. Kruger for his direction of the project and to Mrs. Elizabeth (Lockey) Breining, Miss Alice G. Loe, and Miss Judith Zagrodnick for long-term computations and precision charting. We wish to express our appreciation to Dr. R. B. Murphy for suggestions regarding the new text material and to the Misses Barbara Leetch and Melissa Twigg and to Mr. Spencer W. Roberts for substantial contributions in the later stages of the project. We also wish to thank Miss Ruth L. Stumm and Miss Ernestine F. Manzo for their painstaking secretarial and typing work in the preparation of the text and the index, and the publishers and printers for the care they have exercised in the production of the finished book. Finally we are indebted to Messrs. G. D. Edwards and E. G. D. Paterson, successively Directors of Quality Assurance at Bell Telephone Laboratories, Incorporated, under whose guidance the project was carried on and completed.

H. F. D.
H. G. R.

February 1959

Preface

THE sampling inspection tables presented in this book were developed for use in the manufacture of communication apparatus and equipment for the Bell Telephone System. The tables were published in the January 1941 issue of *The Bell System Technical Journal* and have been adopted for use by several other companies. In the present necessities of wartime production they have been found widely useful and many requests have been received for reprints.

This book has been prepared to meet this demand. It assembles under one cover the three papers of original publication and adds a brief introduction. The papers have been reproduced with no modifications but the material has been rearranged in chapters. Chapter 1 outlines some of the factors to be considered in setting up inspection plans and develops a basis for minimizing the amount of inspection. Chapter 2 covers double sampling, the "average outgoing quality limit" (AOQL) concept, and the mathematical background of the tables. Chapter 3 is a reproduction by permission, for which gratitude is now expressed, of a paper by D. B. Keeling and L. E. Cisne, of the Hawthorne plant of the Western Electric Company, which outlines the shop procedures for applying the tables.

As indicated by the acknowledgments of Chapter 2, many individuals in Bell Telephone Laboratories and in the Western Electric Company contributed to the work underlying the development of these tables. In a cooperative venture of this kind any listing of all of the individuals concerned, and of their specific contributions, is next to impossible. The authors feel, however, that mention should be made of the late Mr. C. N. Frazee, whose pioneering work on the application of statistical methods to inspection problems was a spur to this particular development; of Dr. W. Bartky, for mathematical contributions, including those relating to multiple sampling; of Dr. W. A. Shewhart, whose resourcefulness and companion developments of the theory of "quality control" had a strong influence on the direction in which the work progressed; and of Mr. G. D. Edwards, under whose guidance and supervision the entire program has been carried through.

H. F. D.
H. G. R.

June 1944

Contents

. . .

INTRODUCTION

CHAPTER 1 • A METHOD OF SAMPLING INSPECTION

CHAPTER 2 • SINGLE SAMPLING AND DOUBLE SAMPLING INSPECTION TABLES

CHAPTER 3 · USING DOUBLE SAMPLING INSPECTION IN A MANUFACTURING PLANT

CHAPTER 4 · OPERATING CHARACTERISTICS OF SAMPLING PLANS

Introduction

IT HAS LONG BEEN RECOGNIZED, where sampling instead of complete inspection is used, that certain errors or risks are unavoidable. In the Western Electric Company in 1923 engineers working on the problem of setting up final inspections for customer lots of installed central office equipment used probability theory to provide a quantitative approach. In this work they prepared probability-of-acceptance curves to facilitate a choice between several sampling inspection plans and introduced the concept of a "lot tolerance" for defectives, that is, a limiting percentage of defective articles in a lot. This was followed by intensive developments in a newly formed inspection engineering department in Bell Telephone Laboratories. During the next two years its studies brought forth several concepts and terms which are still in profitable use, notably the "Consumer's Risk" and the "Producer's Risk" associated, respectively, with two specific levels of quality.

The use of these two risks alone for obtaining a sampling plan was insufficient to the needs of the problem. They merely provide the equivalent of a statistical test for a unique sample from a unique lot. What was required was the inclusion of some of the ideas contained in the theory of "quality control" then under development in the Bell Telephone System. These ideas involved the treatment of lots not individually but as a series. Such a treatment provided a basis for securing certain fundamental and over-all advantages by an adjustment of inspection procedure to the actual quality conditions existing at any time. One of the more important practical advantages is "a minimum amount of inspection for a given inspection procedure," which is one of the salient features of the tables in this book.

Shortly thereafter, in 1926, in a program of active cooperation between the Laboratories and the Western Electric Company, the first set of sampling inspection tables was prepared for shop use. These tables provided a complete set of "minimum inspection" solutions for several selected levels of process average quality using the lot tolerance concept with a 10 per cent Consumer's

Risk. Studies had shown, however, that a double sampling procedure offered both theoretical and practical advantages. These tables were accordingly prepared for double sampling as well as single sampling covering ranges of conditions dictated by the needs of actual shop practice.

Experience with the tables indicated some shortcomings in their application to shop inspections. This was particularly true in process and final inspections where "lots" of piece-parts or assemblies were too often merely quantities whose size was determined by convenience in handling. In contrast with customer lots which are specific in quantity and which commonly retain their identity, the inspection lots in manufacture were usually convenient subdivisions of a flow of product. The use of a lot tolerance per cent defective was not too meaningful for such lots. Interest centered much more on controlling the average quality. Accordingly, in 1927, the concept of "Average Outgoing Quality Limit" (designated AOQL) was developed and additional tables were prepared on this basis. The shop inspection organizations found these tables eminently satisfactory, and their use spread rapidly.

The tables now published are the product of several revisions but are based on the same fundamental principles which continue to satisfy engineering needs. They are arranged for convenient use by those concerned with planning inspection, whether incoming inspections of purchased items, process inspections of parts and assemblies, or final inspections of finished product. Their application has contributed in an important way to the reduction in inspection costs and improvement in control of quality.

FACTORS INFLUENCING THE DESIGN OF INSPECTION PLANS

The variety of sampling plans, procedures, and tables that can be constructed is almost unlimited. The advantages and disadvantages of each of several possible choices need to be carefully weighed from both theoretical and practical standpoints. More often than not, the engineering considerations weigh more heavily than the statistical because over-all costs and ease of application under hurried shop conditions are of first importance.

Trials of alternative procedures under actual operating conditions may bring to light unanticipated factors that result in the adoption of a plan which on paper seems less efficient; and the choice of the most advantageous plan is often determined on the proving ground of experience. To cite a specific example, unpublished multiple sampling plans, permitting an unlimited number of samples from a lot, were developed and made available for Bell System use in 1927. Just as double sampling generally required less total inspection than single, so multiple sampling generally required less than double for a given degree of protection, although much of the gain over single sampling resided in the second sample. Now after years of use double sampling seems to be quite generally preferred. This is undoubtedly due to costs both tangible and intangible other than those directly dependent on the number inspected, such as, for example, those associated with the physical selection of samples, score keeping, length of indecision period in doubtful cases, and psychological preferences of inspectors.

Other types of single and double sampling tables have been developed and used; some with concepts differing from lot tolerance and AOQL, others with features to serve particular purposes, for example, in situations where two or more "seriousness classifications of defects" are simultaneously involved. However, in the field in which the authors have been concerned the present tables have continued to be widely used.

APPLICATION OF TABLES

Two general types of tables are given, one based on the concept of lot tolerance and the other on AOQL. The broad conditions under which the different types have been found best adapted are indicated in Chapter 2. For each of the types, tables are provided both for single sampling and for double sampling. Each of the individual tables constitutes a collection of solutions to the problem of minimizing the over-all amount of inspection.

The two features, minimum amount of inspection and AOQL, have been found particularly appropriate in process inspections of piece-parts and subassemblies and in final inspections of finished product manufactured on a quantity basis. With respect to the first feature, the particular "minimum" chosen relates to over-all inspection, including that performed on samples plus that involved in the complete inspection of the remainders of rejected lots. The costs of these two phases of inspection, even when listed separately in cost accounting, are but part of the over-all costs which are to be minimized. For inspections of incoming material, likewise, this same minimum is often of prime importance. For example, for purchased parts, it is often cheaper or more expedient to inspect completely a small proportion of marginally unsatisfactory lots than to reject them on the basis of sampling and then to ship them back to the supplier. The second feature, Average Outgoing Quality Limit, has proved to be not only a convenient and useful concept but one that is understandable to those on the inspection line as well as to management. The gauge of over-all performance is the underlying general level of quality in the product actually delivered. As long as there are adequate safeguards against passing product of abnormally spotty quality, it is reassuring to know that, with properly maintained inspections, the average quality delivered will not be poorer than some definite value. Use of an AOQL inspection plan, with its provisions for minimized inspection, gives such assurance economically.

For process and final inspections in many manufacturing plants, the AOQL double sampling tables have retained first place in general utility. Why double sampling is generally more advantageous than single sampling is considered in Chapter 2. An interesting further reason comes into play when the acceptance number for the sample is necessarily very small. Double sampling always allows one defect, whereas single sampling may allow none. It is a human tendency to avoid taking any important action on the evidence of a single isolated defect. Like a first offense, the single defect may be looked upon lightly. But "two defects" is another story. And double sampling never calls for rejecting a lot on less than two defects.

Another item of practical interest relates to the curtailment of the inspection of a sample when the acceptance number is exceeded before all articles in the sample are inspected. In single sampling this practice is generally avoided for a reason that might not at first be obvious. In any program that is really concerned with control of quality, process and final inspections have a broader objective than merely to serve as an agency for accepting and rejecting individual lots one by one, in effect considering the disposition of each lot as a unique problem. Instead, the results obtained from such inspections can be used to provide a continuing measure of the process average quality, information useful not only to an operating department as a gauge of its performance and as a basis for instituting corrective action but also to the inspector himself for adjusting the procedure in the interests of minimizing his work. Were the inspection to be stopped in single sampling as soon as the acceptance number is exceeded, the data obtained would be biased and ill suited for easy computation of process average quality. For double sampling, the situation is different. It has been the practice to inspect first samples completely, regardless of the findings, in order to provide a continuing check on process average quality. But, where the inspection of the remainder of rejected lots is performed by someone other than the sampling inspector, it has been common practice to stop inspection on a second sample as soon as the acceptance number for the combined samples has been exceeded. This keeps to a minimum the number of articles inspected during sampling. The actual procedure in double sampling should, of course, be taken into account in comparisons of the so-called efficiency of the double sampling plan and other proposed inspection plans where the comparisons relate only to the number inspected in samples.

Whether it is better at a given inspection station to include many inspection characteristics or only a few is a question that involves convenience, economy, and the requirements of one's program of quality control. The advantages of including only a few characteristics at a time, grouping together those subject to like inspection operations, usually outweigh the disadvantages. The inspection procedures are thereby simplified and each characteristic individually receives closer attention and can thus be better controlled. Where, nevertheless, it is desired to include many inspection characteristics at one inspection station, it is sometimes convenient to express quality in terms of defects per unit rather than in terms of per cent defective. In such instances, upon converting the scale of the AOQL and lot tolerance values, the tables become generally applicable. For example, an AOQL of 10 per cent can be regarded as 10 defects per 100 units or 0.1 defect per unit. For this treatment the AOQL values listed in the tables are unaffected, whereas the lot tolerance per cent defective values may be affected in some areas of the tables, but to a degree that is usually of no great practical significance.

The tables have also been found useful as a base for setting up special inspection plans to fit particular conditions. Among these are plans for reduced inspection where several consecutive lots are treated as an inspection lot, with a suitable fraction of the normal sample taken from each presented lot, as long as preceding lots meet a cumulative criterion for control at an extremely good quality level. Also among these are plans for separate treatment of two groupings of defects, such as major defects and total defects, where each grouping has its own AOQL or lot tolerance value. Mathematical relations and charts have been included to assist in obtaining solutions for such special cases.

APPLICATION WHEN REJECTED LOTS ARE NOT SCREENED *

While the tables provide plans that constitute solutions of the problem of "minimum total inspection," questions have arisen regarding whether they may be used in applications where rejected lots are not subjected to 100 per cent inspection, as in destructive inspections and in situations where the producer may or may not do the desired screening.

When destructive inspection is required in order to evaluate the operation or performance of a given product, small samples are generally selected and tested. Where the test is an attributes test, the most common plan is (a) a single sampling plan with an acceptance number, c, of 0, or (b) a double sampling plan with acceptance numbers, c_1 and c_2, equal to 0 and 1. Whether such small sample plans need to be labelled with an LTPD or an AOQL or an AQL † is open to question, but if the plan used is one that appears among the single sampling plans of Appendix 3 or in the single or double sampling AOQL tables of Appendices 6 or 7,‡ the user is able to see how such a plan will operate under various conditions by studying the OC curve of the plan as published herein.

---

* Added in the Second Edition.

† "Acceptable Quality Level."

‡ Single and double sampling plans in the LTPD tables of Appendices 4 and 5 may also be used for these small c, or c_1 and c_2, acceptance numbers. OC curves for single sampling plans are given in Appendix 3. OC curves for a number of LTPD double sampling plans may be estimated from the OC curves of Appendix 2, where the n_1 and n_2 values for an AOQL plan differ only slightly from the n_1 and n_2 values for the LTPD plan.

Speaking more generally, even though rejected lots are not 100 per cent inspected, the AOQL tables "may quite properly be used in the light of the operating characteristics of the plans." * First, product having quality equal to or better than the process average value for the table column used will have a very high probability of acceptance—usually over 0.98 or 0.99, seldom under 0.95. Second, lot quality values for which the probability of acceptance is 0.10 are shown in the tables, and the complete set of Type B operating characteristic curves (OC curves) are shown in Appendices 1 and 2. Third, "product having a per cent defective moderately higher than, say, one and one half times the AOQL value of the plan used will be subject to substantial rejections, commonly sufficient to force corrective action by the producer. For example, for the 2 per cent AOQL double sampling tables, and for a product per cent defective one and one half times the AOQL value, there will generally be 15 per cent to 30 per cent rejection for lot sizes under 1,000 and 25 per cent to 40 per cent for lot sizes over 1,000. Thus, while there will be no guarantee that the average outgoing quality will be better than the AOQL value if rejected lots are not inspected 100 per cent, experience indicates that the plan, through its power of rejection, will tend to compel the producer, in his own interests, to maintain a process quality level which at worst will be little, if any, poorer than the AOQL. The pressure will generally be greater, the larger the lot size." †

CHOOSING A SAMPLING PLAN ‡

As the engineer goes about choosing a sampling plan for a particular application, he must make a number of decisions which depend on the conditions under which the plan is to be used. The accompanying "Sequence of Steps" gives an outline of a typical procedure. These steps are shown also in the following numbered paragraphs, together with references to places in the text where the reader will find discussions or comments that may assist him in his decisions. To provide further assistance, some additional thoughts are provided in supplementary lettered subparagraphs for the several steps.

* Harold F. Dodge, "Administration of a Sampling Inspection Plan," *Industrial Quality Control*, Vol. V, No. 3, November 1948, pp. 18–19.

† *Ibid.*

‡ Added in the Second Edition.

Sequence of Steps

1. DECIDE what characteristics to include.

2. DECIDE what is to constitute a lot.

3. CHOOSE the type of protection— LTPD or AOQL.

4. CHOOSE a suitable value of LTPD or AOQL.

5. CHOOSE between single sampling and double sampling.

6. SELECT the appropriate sampling table in Appendix 4, 5, 6, or 7.

7. OBTAIN an estimate of the PA (process average per cent defective).

8. CHOOSE a sampling plan for the given lot size and estimated PA.

9. FIND the OC curve of the sampling plan. If it is satisfactory, adopt the plan.

10. SELECT sample units from the lot by a random procedure.

11. FOLLOW the prescribed procedure — single sampling or double sampling.

12. KEEP a running check of the PA. Change the sampling plan as necessary to match shifts in the PA.

1. *Decide what characteristics will be included in the inspection* (pp. 3, 46).

   (a) If advantageous, use a separate sampling plan for a single characteristic or a selected group of characteristics of like seriousness. Sampling need not wait until all characteristics have good quality.

   (b) If one or two characteristics give an outstanding number of defects, treat them separately (sampling or 100 per cent inspection; also, if possible, concentrate on correcting the causes of trouble) and include the rest collectively in the sampling inspection.

(c) If all characteristics have satisfactory quality, include all of them collectively in the sampling inspection.

(d) In general, combine at one inspection station characteristics subject to essentially similar inspection operations, e.g., all visual inspection items together, all gauging, or all testing. Visual and gauging inspection operations often combine well.

(e) Include in any one group characteristics of essentially the same degree of seriousness. [If two degrees of seriousness are involved, say major and minor, keep all majors together in one group and all minors in a second group, and treat them as indicated in Item 8(e) on page 7.]

(f) Consider these plans applicable to all basic types of inspection for manufactured products—receiving, process, and final—and to the inspection of administrative and clerical products as in "paper-work quality control."

2. *Decide what is to constitute a lot for purposes of sampling inspection* (pp. 21, 45).

(a) So far as practicable, require that individual lots presented for acceptance comprise essentially homogeneous material from a common source.

(b) If presented material comes from two or more distinct sources not under a common system of control, arrange to have each presented lot comprise material from only one of those sources; otherwise have source identification information furnished with each lot.

(c) To minimize the amount of inspection, make the lots as large as practicable, considering the limitations of available storage space, delays in shipment, difficulty in handling large rejected lots, etc.

3. *Choose between lot quality* (LTPD) *and average outgoing quality* (AOQL) *protection* (pp. 2, 5, 24, 30).

(a) Choose AOQL if interest centers on the general level of quality of product *after* inspection. AOQL plans have been found generally more useful than LTPD plans in inspections of a continuing supply of product, especially in consumer's acceptance inspections and in producer's receiving, process, and final inspections.

(b) Choose AOQL if you want a quality figure that will almost invariably be safely met by the running average quality of product after inspection.

(c) Choose LTPD if you want a quality figure that will almost surely be met by every lot. (This will be a much more pessimistic figure than the AOQL value of the plan.)

(d) As a manufacturer trying to meet a consumer's stated AQL,* use for final inspection an AOQL plan with an AOQL value equal to the specified AQL value, in order to provide good assurance that outgoing quality will be found acceptable by the consumer (or set the AOQL at one and one third times the AQL for reasonably good assurance).

(e) When producer and consumer of a product are two departments of the same company, use AOQL plans with the provision that the producer perform the 100 per cent inspection of rejected lots. Close interchange of quality findings will expedite good process control of quality.

(f) Wherever practicable, make arrangements for the producer to perform the 100 per cent inspection of rejected lots under procedures acceptable to the consumer and to provide suitable certifications of work performed.

4. *Choose a suitable figure of quality* (LTPD *or* AOQL) *for the sampling plan to be used* (pp. 6, 46, 52–53).

(a) For LTPD, choose the value of per cent defective that you are willing to accept not more than 10 per cent of the time (i.e., reject at least 90 per cent of the time).

* AQL = Acceptable Quality Level, as used to index certain systems of sampling plans, signifying what the consumer feels to be the maximum per cent defective (or the maximum number of defects per 100 units) that can be considered satisfactory as a process average. See H. A. Freeman, M. Friedman, F. Mosteller, and W. A. Wallis, *Sampling Inspection*, McGraw-Hill Book Co., Inc., New York, 1948, and *MIL-STD-105A, Sampling Procedures and Tables for Inspection by Attributes*, Supt. of Documents, Govt. Printing Office, Washington, D. C., 1950.

(b) For AOQL, choose the value of *average per cent defective in product after inspection* that should not be exceeded.

(c) In choosing a value of LTPD (or AOQL), consider and compare the cost of inspection with the economic loss that would ensue if quality as bad as the LTPD were accepted often (or if the *average level* of quality were greater than the AOQL). Even though the evaluation of economic loss may be difficult, relative values for different levels of per cent defective may often be determined.

(d) As a guide to the choice of AOQL values, note the examples * (in the accompanying table) used in one shop under prescribed conditions.

* Harold F. Dodge, "Administration of a Sampling Inspection Plan," *Industrial Quality Control*, Vol. V, No. 3, November 1948. The examples include those given on p. 46 of this book, as well as others furnished by Mr. G. D. Milne, Western Electric Company.

5. *Choose between single sampling and double sampling* (pp. 30–31).

(a) In general, for economy in over-all inspection effort, use double sampling rather than single sampling.

(b) In general, for minimum variation in the inspector's workload, use single sampling.

(c) Consider adopting double sampling as the normal standard for sampling plans in a given plant, with a view to effecting over-all economies. Double sampling has been adopted as standard in many areas for this reason.

(d) In a particular case, for a given AOQL and given process average, compare [per item 9(b)] the OC curves of the two sampling plans (single sampling and double sampling) as an aid in making a choice.

6. *Select the proper sampling table in Appendix 4, 5, 6, or 7 on the basis of the above choices.*

EXAMPLES OF AOQL VALUES CHOSEN FOR VARIOUS INSPECTIONS

Description of Units	Type of Inspection	AOQL Value
Metallic and nonmetallic rods, tubes, strips, sheets	Visual, dimensional	3.0%
Milled parts	Visual, dimensional	0.75%
Molded plastic parts	Visual, dimensional	1.0%
Die-cast parts	Visual, dimensional	1.0%
Formed and drawn parts	Visual, dimensional	2.0%
Castings, ferrous and nonferrous	Visual, dimensional	3.0%
Machine screws	Five dimensions	2.0%
Hexagon nuts	Visual inspection after zinc plating	2.0%
Twin eyelets	Six dimensions and 4 visual requirements	3.0%
Ceramic insulators	Visual, dimensional	5.0%
Miscellaneous inexpensive electrical apparatus	Breakdown	0.25%
Miscellaneous completed electrical apparatus	Resistance	0.5%
Relay coils *	Inductance and electrical breakdown	1.0%

* This is a process check for the specified requirements; it is supplemented by another sampling inspection after assembly.

7. *Obtain an estimate of process average per cent defective* (pp. 13, 31–32, 47, 53).

    (a) Use immediately prior data to estimate the process average.

    (b) Use rough estimates at the start, if little or no actual data are available; a poor estimate merely prevents getting the most economical plan but keeps the same (LTPD or AOQL) protection.

    (c) As more data are collected, make improved estimates of process average.

    (d) Omit wild and obviously nonrepresentative sets of data in making estimates and adopt some suitable rule for discarding data (see footnote, p. 32).

8. *Choose a sampling plan for the given lot size and the estimated process average* (pp. 26, 32, 47).

    (a) If the estimated process average per cent defective (hereinafter designated PA) falls within the range of PA values in the selected table, choose the sampling plan corresponding to the PA value and to the given lot size.

    (b) If the PA is unknown or is estimated to be larger than the largest PA value given in the table, choose the sampling plan corresponding to the largest PA in the table (last column) and to the given lot size.

    (c) Under (b), obtain revised estimates of the PA from the lot-by-lot data and use a sampling plan with a smaller sample size as soon as a revised estimate of the PA permits.

    (d) If, for single sampling, the sampling plan given by the table has $c = 0$, consider whether it would be preferable to use a plan with $c = 1$ to avoid making rejections on finding a single defective. There is no such problem for double sampling, since c_2 always equals 1 or more.

    (e) If inspection includes two classes of defects, major and minor, with two AOQL values, choose the two sampling plans from the appropriate tables in the Appendices and use them simultaneously. If, however, it is desired to use the *same sample size* for both *majors and minors*, proceed as follows:

        (1) Choose the sampling plan for major defects from the appropriate table in the Appendices.

        (2) If this is a single sampling plan, then for minor defects use a single sampling plan with the same sample size, n, but a value of c corresponding to the AOQL for minors. To find the value of c, compute

$$y_e = \text{AOQL} \left(\frac{Nn}{N - n} \right),$$

or, if $n/N < 0.05$,

$$y_e = \text{AOQL} \ (n).$$

Look in Table 2–3 for the largest y value that is less than y_e and use the corresponding value of c.*

        (3) If a double sampling plan is used for majors, use *single sampling* for minors with the sample size, n, equal to the first sample size, n_1, of the double sampling plan and a value of c corresponding to the AOQL for minors; see subparagraph (2) above.*

9. *Find the OC curve of the chosen plan (or plans) in the Appendices. If the plan appears satisfactory, adopt it for use.*

    (a) For AOQL plans, find the OC curve in Appendix 1 (single sampling) or Appendix 2 (double sampling). For LTPD plans, see pp. 55–56 for the procedure to get the OC curve.

    (b) If two plans (single sampling and double sampling) were tentatively selected per paragraph 5(d), compare their OC curves as an aid to a final choice.

10. *From the lot, select sample units by means of a random procedure* (p. 23).

    (a) Consider the use of random numbers as the preferred way of selecting sample units "at random." Each unit in the lot is assigned a serial number (usually on paper), and then those units whose serial numbers correspond to the numbers in some section of a table of random numbers are included in the sample.

---

* The solutions of subparagraphs (2) and (3) presume the usual condition that the AOQL for minors is a larger per cent defective than the AOQL for majors. It provides a minimum total inspection for majors, but, instead of providing minimum total inspection for minors, substitutes the convenience of the same sample size for minors as for majors.

(b) If a double sampling plan has been chosen, consider selecting sample units for both samples at the same time.

11. *Follow the sampling inspection procedure for single sampling or double sampling, whichever was chosen* (pp. 23–24).

    (a) Inspect each unit in the sample for all the characteristics decided on in step 1.

    (b) If single sampling is being used, inspect all units in the sample even though the acceptance number is exceeded before all units have been inspected. This facilitates estimation of the process average.

    (c) If double sampling is being used, inspect all units in the first sample; if desired, discontinue inspection of the second sample when the acceptance number, c_2, is exceeded.

12. *Keep a running check on the process average and change the sampling plan if the process average changes sufficiently* (pp. 47, 53).

    (a) Adopt a definite plan for making periodic estimates of the process average—every 20 or 50 lots or every month, quarter, or six months, depending on the production rate and the quality history.

    (b) Keep the producing organization informed of the running quality of presented product, preferably in control-chart form, and furnish prompt information regarding any sudden adverse shifts in quality.

    (c) Change from one sampling plan to another within a sampling table, as the process average changes from one general level to another. This provides a general basis for tightened and reduced inspection while holding to a given AOQL. If, with stable quality at an excellent level, it is desired to reduce inspection even further, use a larger AOQL value, say twice as large as the basic AOQL.

Chapter 1

A Method of Sampling Inspection

This chapter outlines some of the general considerations which must be taken into account in setting up any practical sampling inspection plan. An economical method of inspection is developed in detail for the case where the purpose of the inspection is to determine the acceptability of discrete lots of a product submitted by a producer. By employing probability theory, the method places a definite barrier in the path of material of defective quality and gives this protection to the consumer with a minimum of inspection expense.

The material in this chapter was originally published in *The Bell Syste m Technical Journal*, Vol. VIII, No. 4 (October 1929), pages 613–631. Minor revisions only have been made for this edition.

ONE OF THE COMMON QUESTIONS in everyday inspection work is, *How Much Inspection?* The answer must always be arrived at in the light of economy, for only the least amount of inspection which will accomplish the purpose can be justified.

We wish here to consider the problem of setting up an economical inspection plan whose immediate purpose is the elimination of individual lots of product which are unsatisfactory in quality. By a lot of unsatisfactory quality is meant one that contains more than a specified proportion of defective pieces. This proportion is usually small and is based on economic considerations. The interest in inspections made for the elimination of such lots is shared by two parties—the producer and the consumer. The consumer establishes certain requirements for the quality of delivered product. The producer so arranges his manufacturing processes and provides such an inspection routine as will ensure the quality demanded. Our problem recognizes that the consumer runs some risk of receiving lots of unsatisfactory quality, if the quality of each lot is judged by the results of inspecting only a sample. The method of attack presumes the adoption of a risk whose magnitude is agreeable to the consumer and the selection of a particular inspection procedure which will involve a minimum of inspection expense while guaranteeing the protection agreed upon.

Considering various possible concepts of a risk, the consumer would prefer to have one adopted that was worded as follows:

Not more than a specified proportion of the delivered lots shall be unsatisfactory in quality.

In other words, he would like to have the assurance that the risk of receiving a lot of unsatisfactory quality shall not exceed some definite figure. This risk involves two probabilities:

(a) that an unsatisfactory lot will be submitted for inspection, and

(b) that the inspector will pass as satisfactory an unsatisfactory lot submitted for inspection.

Without definite information regarding probable variations in the producer's performance and without assurance that this performance will remain consistently the same, the use of probability (a) in stating the risk to the consumer might be misleading. This probability will therefore receive no further consideration in this chapter so far as the definition of the risk to the consumer is involved.

For probability (b) a reasonably low value for an upper limit can be given without any knowledge of the producer's performance. This upper limit (as defined on p. 12) has been taken as the starting point for the inspection method presented in this chapter. The concept of risk which has been adopted gives the consumer unconditional assurance that his chance of getting any unsatisfactory lot *submitted for inspection* shall not exceed some definite magnitude. This magnitude is the Consumer's Risk used.

Our problem may hence be stated as follows:

Given a product of a specified type of apparatus or material coming from a producer in discrete lots, what inspection plan will involve a mini-

mum of inspection expense and at the same time ensure that under no conditions will more than a specified proportion of the unsatisfactory lots submitted for inspection be passed for delivery to consumers?

This problem is economic in character and its solution involves the use of probability theory for establishing the height of the barrier to be placed in the path of unsatisfactory material.

Our attention will be directed particularly to the inspection of material which is produced more or less continuously on a quantity basis as distinguished from intermittent production in relatively small amounts. Under these conditions it is possible to secure a continuing record of performance and to set up an inspection program which takes advantage of current quality trends.

1.1 GENERAL CONSIDERATIONS IN SETTING UP AN INSPECTION METHOD

The broad purpose of inspection is to control quality by critical examinations at strategic points in the production process. Raw materials must be inspected. Some of the rough and finished parts must be inspected. In the manufacture of even the simpler kinds of merchandise, inspections dot the chart of progress from raw materials to the finished product. The distribution of inspection activities throughout any process must be so ordered that the net cost of production will be consistent with the quality demanded by the customer. To determine whether an inspection should be made, or how much should be made at any one of the formative stages, is a major problem involving questions of both quality and economy.

One hundred per cent inspection is often uneconomical at a point in the production process where inspection is clearly warranted, particularly when preceded or followed by other inspections, inasmuch as the cost of more inspection at that point may not be reflected by a corresponding increase in the value of the finished product. In special cases, for example where inspection is destructive, 100 per cent inspection may be totally impracticable. Sampling inspections are often best from the standpoint of both the producer and the consumer when the value of quality and the cost of quality are weighed in the balance.

To arrive at an answer to the question, *How Much Inspection?* it is first essential to define clearly just what the inspection is intended to ac-

complish and to weigh all of the important factors both preceding and following the inspection in question which have a direct influence on the quality of the finished product and which as a whole determine how large a part this inspection step must play in controlling quality. Should it serve as an agency for making sure that the product at this stage conforms 100 per cent with the requirements for the features inspected? If so, 100 per cent inspection is required.* Or should it serve to make reasonably certain that the quality passing to the next stage is such that no extraordinary effort would have to be expended on defective material? If so, sampling inspection may be employed. Ahead of all else, decisions are needed as to the specific requirements that must be satisfied by the inspection plan itself. This part of the problem is a practical one—one which must be approached in the light of experience, with a knowledge of present conditions and the statistics of past performance. Once the basic requirements of the plan are agreed upon, probability theory can assist in formulating the details which will accomplish the desired results. It is important to hold in mind that statistical methods are aids to engineering judgment and not a substitute for it.

An attempt has been made in Fig. 1–1 to show schematically some of the outstanding general considerations which must be taken into account in establishing a proper setting for any problem that seeks to determine the economical amount of inspection. Inspections vary widely in purpose, type, and character. While their broad purpose is to control quality, the immediate objectives of individual inspection steps differ. For example, the objective of one step may be to secure information which will assist directly in controlling the manufacturing process by detecting errors or trends in performance which would become troublesome if allowed to persist unchecked. In other places the immediate objective may be to determine the acceptability of definite quantities of product or to provide a screen for sorting the bad pieces or the bad lots from the good ones. Materials, parts in process, and finished units are scrutinized with these objectives in view. Depending on circumstances, the character and completeness of inspections vary from visual examination of small samples to careful measurement or testing of each piece.

---

* Practically, of course, 100 per cent conformance may not be attainable because of lack of 100 per cent efficiency in the inspection operation itself.

1.2 CONDITIONS UNDER WHICH THE PRESENT METHOD APPLIES

A large amount of industrial inspection work consists in comparing individual pieces with a standard—such as gauging a dimension or measuring an electrical property—to determine whether the pieces do or do not conform with the requirements given in specifications. This is often referred to as inspection by the "method of attributes." Consideration will be directed here to the case where nondestructive sampling inspection of this kind is conducted on discrete lots of product for the purpose of determining their acceptability.

From the standpoint of sampling theory, one of the general requirements is that each lot should be composed of pieces which were produced under the same essential conditions. Practically, this means that an attempt should be made to avoid grouping together batches of material, which, due to manufacturing conditions or methods, are apt to differ in quality. It is presumed, of course, that the sample drawn from any lot will be a random sample so that it may fairly represent the quality of the entire lot.

Summing up the general conditions for which a solution is sought, we assume

(1) The purpose of the inspection is to determine the acceptability of individual lots submitted for inspection, i.e., sorting good lots from bad.

(2) The inspection is made by the "method of attributes" to determine conformance with a particular requirement, i.e., each piece does or does not meet the limits specified.

(3) A lot is homogeneous in quality and the sample from it is a random sample.

The items marked with the symbol ⊕ in Fig. 1–1 indicate the set of conditions involved in our problem.

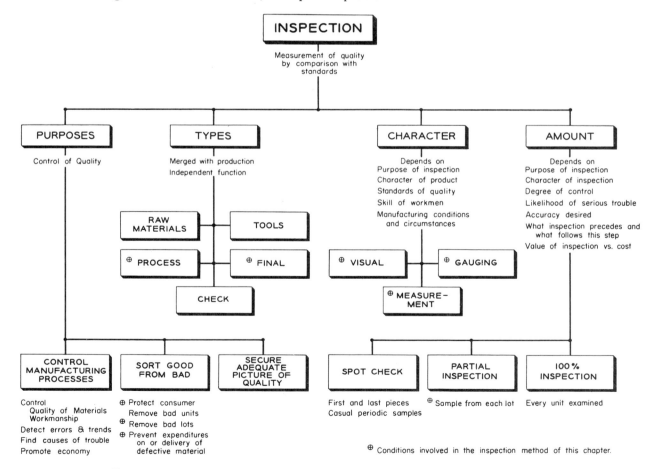

Fig. 1–1   Some of the factors to be considered in establishing an inspection plan

1.3 PROTECTION AND ECONOMY FEATURES OF THE METHOD

The adoption of sampling inspection at any stage of manufacture carries with it the premise that the product emerging from this point does not have to conform 100 per cent with specification requirements. It is often more economical, all things considered, to allow a small percentage of defective pieces to pass on to subsequent assembly stages or inspections for later rejection than to bear the expense of a 100 per cent inspection. Under these conditions, the status of the inspection can be clarified by establishing a definite tolerance for defective pieces (hereinafter called *defectives*) in the lots submitted to the inspector for acceptance. This may be specified as a limiting percentage defective, a figure which may be considered as the border line of distinction between a satisfactory lot and an unsatisfactory one. Thus, if the percentage defective is greater than this "tolerance per cent defective," the lot is unsatisfactory and should be rejected. We say "should be" rejected, but this cannot be accomplished with absolute certainty if only a sample is examined. Sampling inspection involves taking chances, since the exact quality of a lot is not known when only a part is inspected. According to the laws of chance, a sample will occasionally give favorable indications for bad lots which will result in passing them for delivery to consumers.

The first requirement for the method will therefore be in the form of a definite assurance against passing any unsatisfactory lot that is submitted for inspection.

The second requirement that will be imposed is that the inspection expense be a minimum, subject to the degree of protection afforded by the first requirement.

For the first requirement, there must be specified at the outset a value for the tolerance per cent defective as well as a limit to the chance of accepting any submitted lot of unsatisfactory quality. The latter has, for convenience, been termed the Consumer's Risk and is defined, numerically, as the probability of passing any lot submitted for inspection which contains the tolerance number of defectives.

As will be shown further on, the first requirement can be satisfied with a large number of different combinations of sample sizes and acceptance criteria. To satisfy the second requirement, it is necessary then to determine the expected amount of inspection for a variety of inspection plans, determine the cost of examining or testing, add the costs other than those incurred in the simple process of examining samples, and choose among these plans that which involves a minimum of inspection expense.

There are, of course, a number of possible general methods of inspection procedure, such as single sampling, double sampling, multiple sampling, etc., which allow the examination of only one sample, of two samples, or of more than two samples before a prescribed disposition of the entire lot is made. For each of these general methods, different combinations of sample sizes and acceptance criteria can be found which will satisfy the first requirement. We now prescribe that any lot which fails to pass the sampling requirements shall be completely inspected. Under this condition, one of the above-mentioned combinations will give a lesser amount of inspection than the rest. Since a major cost item is that associated with the *amount* of inspection, we will carry through in detail the problem of finding the combination which will result in the *minimum* amount of inspection for one simple general method of inspection.

1.4 SINGLE SAMPLING METHOD OF INSPECTION

Attention is now directed to what is termed the "Single Sampling" method of inspection, which involves the following procedure:

(a) Inspect a sample.

(b) If the acceptance number for the sample is not exceeded, accept the lot.

(c) If the acceptance number is exceeded, inspect the remainder of the lot.

The term "Acceptance Number" is introduced to designate the allowable number of defectives in the sample.

For this procedure, the first requirement reduces the problem to one which can be solved readily by determining probabilities associated with sampling from a finite lot containing the tolerance number of defectives. For any sample size, there is a definite probability of finding no defectives, of finding exactly one defective, exactly two defectives, etc. If, under the above conditions, the acceptance number were 1, for example, there is one value of sample size such that the probability of finding one or less defectives is equal to the value of the Consumer's Risk specified. Since a lot will be accepted if the observed number of defectives does not exceed the acceptance number, the probability of finding one or less defectives in a sample selected from a lot of tolerance quality is the risk of accepting any lot of tolerance quality submitted to the inspector. It follows that the risk of accepting a lot of worse-than-tolerance quality is less than the Consumer's Risk just defined. If the producer gets into trouble and begins to submit lots of unsatisfactory quality, the consumer has the assurance that his chance of getting them will not exceed this figure. In fact, the worse the quality, the less will be their chance of passing without a detailed inspection. Thus the amount of inspection is automatically increased as quality degenerates.

For every acceptance number, such as 0, 1, 2, etc., there is a unique size of sample which will satisfy the specified values of tolerance per cent defective and Consumer's Risk. We thus have many pairs of values of sample size and acceptance number from which to choose.

The second requirement dictates which pair shall be chosen. We will select that pair which involves the least amount of inspection for product of *expected* quality. In industry the quality emerging from any process tends to settle down to some level which may be expected more or less regularly day by day. If this level could be maintained quite constant, if the variations in quality were no larger than the variations that could be attributed to chance, then inspection could often be safely dispensed with. But practically, while such a level may be adhered to most of the time, instances of man failure or machine failure are bound to arise spasmodically, and as a result the quality of the output may gradually or suddenly become unsatisfactory. The method of solution takes into consideration this usual or expected quality and requires an estimate of the expected quality under normal conditions. A satisfactory estimate of this can usually be obtained by reviewing data for a past period during which normal conditions existed and by utilizing such other pertinent information as bears on manufacturing performance under present or anticipated conditions. This expected value is defined as the *process average* to be used in the solution. Thus the method under discussion will assure the producer of a minimum of inspection expense as long as he holds to his expected performance. If he gets into trouble and the quality becomes poorer than normally expected, the method automatically increases the inspection by an amount which varies with the degree of quality degeneration. This reacts on the producer directly by increasing his inspection expense and

serves as an incentive to the elimination of the causes of trouble. The producer's expected performance is thus made use of in a way that affects the economy of the inspection work but is not used to color or affect the magnitude of the Consumer's Risk.

The amount of inspection that will be done in the long run for uniform product * of process average quality is made up of two parts:

(1) The number of pieces inspected in the samples.

(2) The number of pieces inspected in the remainder of those lots which fail to be accepted when a sample is examined.

But what proportion of the lots will fail to be accepted on the basis of the sampling results? Here is where probability theory comes in again. There will be a definite probability of exceeding the acceptance number in samples drawn from material of process average quality.† Since we are interested in the amount of inspection *in the long run*, the sample at this stage of the problem may .be regarded as drawn from a very large (mathematically infinite) quantity of homogeneous product whose percentage defective is equal to the process average per cent defective. Thus, for example, with an acceptance number of 1, the average ‡ number of pieces inspected *per lot* as a result of *extended* inspections is equal to the number of pieces in the remainder of a lot multiplied by the probability of finding more than one defective in a sample drawn from an infinite quantity of material of this quality. This value plus the number of pieces inspected in the sample gives the average amount of inspection per lot for an acceptance number of 1. Similar results are found for all other acceptance numbers, and the desired solution is obtained by choosing that acceptance number for which the average amount of inspection per lot is a minimum.

The plan thus provides a definite inspection routine to follow, such that the total inspection effort will be a minimum under normal conditions.

---

* By "uniform product" is meant one produced under a constant system of chance causes, giving rise to a quality which is a chance variable. In the present case this chance variable is assumed to be binomially distributed.

† See Appendices 1, 2, and 3 for curves of operating characteristics (OC curves) for specific sampling plans.

‡ It is to be noted that wherever "average" appears in this chapter "expected" value in the rigorous probability sense is meant.

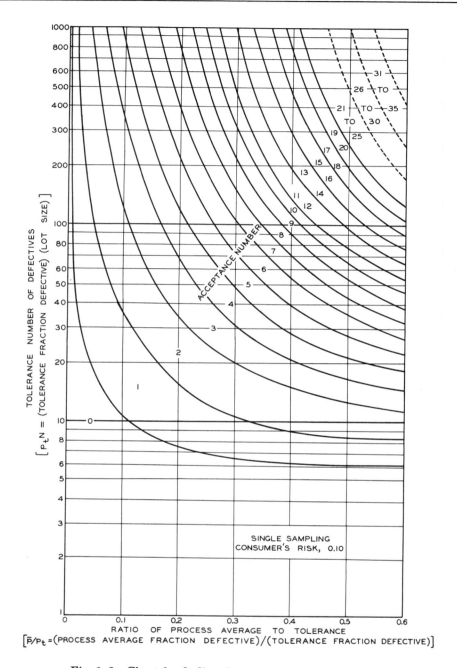

Fig. 1–2   Chart for finding the acceptance number

1.5   CHARTS FOR SINGLE SAMPLING

For any specified value of Consumer's Risk, charts may readily be constructed to give the acceptance number, the sample size, and the average number of pieces inspected per lot for the conditions outlined above. To illustrate the general character of these charts, Figs. 1–2, 1–3, and 1–4 are presented for a Consumer's Risk value of 10 per cent.

In Section 1.8 it is shown that the acceptance number which satisfies the condition of minimum inspection is dependent on two factors: (1), the tolerance number of defectives for a lot, and (2), the ratio of process average to tolerance. Figure 1–2, based on this relationship, defines *zones* of acceptance numbers for which the inspection is a minimum.

Figure 1–3 gives curves for finding the sample size. The mathematical basis for these curves is likewise given in Section 1.8. For a given tolerance number of defectives and the acceptance number found from Fig. 1–2, the value of tolerance times sample size is determined. This quantity divided by the tolerance gives the sample size. The curves shown are based on an approximation which is satisfactory for practical use when the tolerance for defectives does not exceed 10 per cent and the sample size is not extremely small.

Figure 1–4 gives curves which enable one to determine the minimum average number of pieces inspected per lot for uniform product of process average quality. For a given tolerance number of defectives and a given ratio of process average to tolerance, a value of tolerance times minimum average number inspected per lot is determined. This when divided by the tolerance gives the desired value of minimum average number inspected per lot.

1.6 ILLUSTRATIVE EXAMPLE

Suppose that lots of 1,000 pieces each are inspected for a characteristic having a specified tolerance of 5 per cent defective and that the process average quality of submitted lots is 1 per cent defective. If it is desired to have a risk of 10 per cent of accepting a 5 per cent defective lot, what single sampling plan should be followed by the inspector to give a minimum amount of inspection and how much inspection per lot will be required on the average?

Referring to Fig. 1–2 for "Tolerance Number of Defectives" $(0.05 \times 1,000) = 50$ and "Ratio of Process Average to Tolerance" $(0.01/0.05) = 0.20$, we find the acceptance number = 3.

Referring to Fig. 1–3 for "Tolerance Number of Defectives" = 50 and "Acceptance Number" = 3, we find "Tolerance Times Sample Size" = 6.5. Dividing by the tolerance (expressed as a fraction defective) = 0.05, gives a sample size of 130.

Referring to Fig. 1–4 for "Tolerance Number

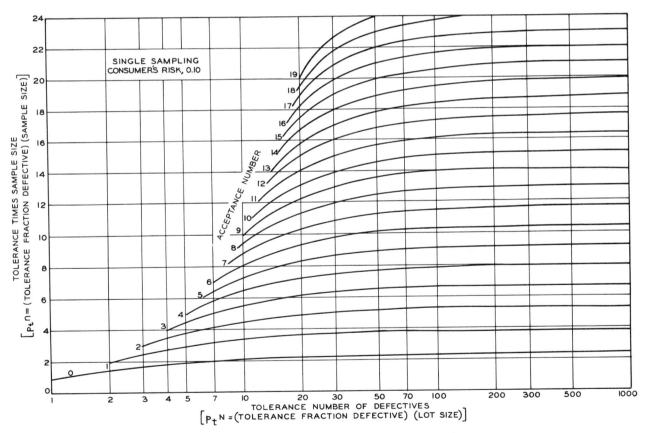

Fig. 1–3 Curves for finding the size of the sample

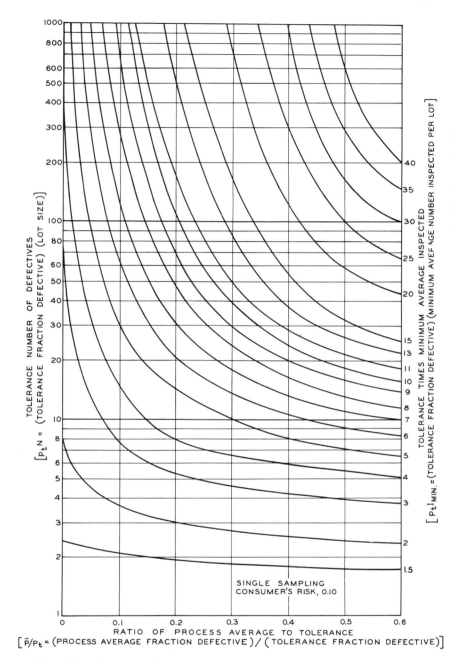

Fig. 1–4 Curves for finding the minimum amount of inspection per lot

of Defectives" = 50 and "Ratio of Process Average to Tolerance" = 0.20, we find by interpolation that the "Tolerance Times Minimum Average Number Inspected" = 8.2. Dividing by the tolerance, 0.05, gives an average number inspected per lot of 164.

This solution has thus been obtained by the initial specification of, first, the tolerance per cent

defective for a single lot and a value for the Consumer's Risk, these two factors being combined to give a definite measure of protection against passing faulty material, and second, a minimum amount of inspection for product of process average quality.

These two requirements control the exact details of inspection procedure and must be initially chosen on the basis of practical considerations and circumstances. The specification of these factors lends definiteness to the problem of inspection and provides a rationalized basis of procedure which can be depended on to give the desired degree of protection. Obviously, any value of Consumer's Risk may be chosen according to circumstances. The value which is proper in any case is dependent on the conditions associated with the product inspected, such as the degree of control exerted by the producing agencies or the protective measures which precede or follow the inspection step in question. For any value of Consumer's Risk, sets of charts similar to Figs. 1-2, 1-3, and 1-4 may be constructed.

1.7   OTHER METHODS OF INSPECTION

The above detailed discussion has been limited to one simple method of inspection, single sampling, in order to show how certain concepts and principles may be applied with the aid of sampling theory. The same principles are readily extended to a plan of double sampling or of multiple sampling, sampling plans in which a second sample may be examined if the first fails or a third, fourth, etc., examined if the preceding samples fail, before resorting to a detailed inspection of the remainder of a lot. As a matter of fact, for given values of tolerance and risk, the minimum average amount of inspection per lot will be somewhat less for plans which permit the examination of more than one sample before detailing; but when consideration is given to the costs associated with interruption of work, extraction of additional random samples, inconveniences or difficulties in handling the routine called for, etc., it has not been found economical in general to examine more than two samples from any lot.

It may be well to point out that other basically different requirements may be chosen for setting up economical sampling inspection plans. For example, a satisfactory method has been devised to meet the following requirements: *

(1) A limiting value to the average per cent defective after inspection.

(2) A minimum amount of inspection for product of process average quality.

This method has been found of value in continuous production where the inspection is in-

* The basis of AOQL sampling plans; see Chapter 2.

tended to serve as a partial screen for defective units. It differs from that described above in that it provides a fixed limit to the *average* quality of product after inspection rather than a limit to the quality of *each individual lot.*

The solution of such problems, which employ probability theory as an aid, always demands a concise statement of the conditions and the specification of numerical requirements which the inspection must satisfy.

1.8   MATHEMATICAL RELATIONS

The problem considered is to minimize the average number of pieces inspected per lot in Single Sampling Inspection. The method and equations developed below may be extended to problems involving 2, 3, 4, etc., samples. The method of extension is somewhat complicated, although the procedure is identical in nature. For example, in Minimum Double Sampling, first and second acceptance numbers and corresponding first and second sample sizes must be found, and at the same time the total Consumer's Risk must be properly divided between the two samples. We will restrict our attention here to the problem of Minimum Single Sampling.

The first seven variables defined below under "Nomenclature" enter into the two equations needed for the solution of the problem, the first three being fixed by the requirement of the method that a definite protection be provided against accepting faulty material, the fourth being fixed by the requirement that the average amount of inspection shall be a minimum for uniform product of process average quality. Therefore, the three unknown variables are c, n, and I. The five variables N, n, I, $\bar{p}$, and p_t are replaced in the solution by four variables which have been obtained from the original variables by combining p_t with the other four, viz., $M = p_t N$, $a = p_t n$, $z = p_t I$, and $k = \bar{p}/p_t$. Since M, P, and k are specified by the method, the unknown variables are c, a, and z. The disposition of the variables is indicated in Table 1-1.

Nomenclature

N = number of pieces in the lot

P = Consumer's Risk: the probability of accepting a submitted lot of tolerance quality

p_t = lot tolerance fraction defective

Table 1–1

DISPOSITION OF VARIABLES

Initial Variables Involved	Initial Fixed Variables	Initial Unknown Variables	Variables Used in Method	Fixed Variables of Method	Unknown Variables of Method
N	N		$M = p_t N$	M	
P	P		P	P	
p_t	p_t		$k = \bar{p}/p_t$	k	
$\bar{p}$	$\bar{p}$		c		c
c		c	$a = p_t n$		a
n		n	$z = p_t I$		z
I		I			

$\bar{p}$ = process average (expected) fraction defective * in the submitted product

c = acceptance number: the stated allowable number of defectives in a sample of stated size

n = number of pieces in the sample

I = average (expected) number of pieces inspected per lot

$M = p_t N$ = number of defectives in a lot of tolerance (p_t) quality

$a = p_t n$ = expected number of defectives in a sample drawn from a lot of tolerance (p_t) quality

$z = p_t I$ = product of the tolerance and the average (expected) number of pieces inspected per lot

$k = \bar{p}/p_t$ = ratio of process average fraction defective to tolerance fraction defective

m = number of defectives found in the sample

$C_n^N = \dfrac{N!}{(N-n)!\,n!}$ = number of combinations of N things taken n at a time

The solution of the problem requires the consideration of the following two equations:

* The symbol $\bar{p}$, as used in this problem, is an assumed true parameter of the universe sampled, and according to the notation adopted by Bell Telephone Laboratories should be primed, i.e., $\bar{p}'$. For the sake of simplicity here the prime notation has been omitted in the equations. Reference, W. A. Shewhart, "Quality Control," *The Bell System Technical Journal*, Vol. VI, No. 4 (October 1927), p. 723, footnote 3.

$$z = f_1(M, a, c, k), \qquad (1\text{--}1)$$

$$P = f_2(M, a, c), \qquad (1\text{--}2)$$

where f_1 and f_2 represent symbolic functions which are to be determined later.

We wish to find a pair of values (c, a) which will make z a minimum, subject to the condition that this pair (c, a) satisfies equation (1–2). Hence, due to the discreteness of c, pairs (c, a) satisfying (1–2) are substituted in (1–1) until a minimum value of z is found. Thus, for $P = 0.10$ and for given values of M and k, we read c from Fig. 1–2, obtain the value of a for this value of c from Fig. 1–3, and determine the minimum value of z from Fig. 1–4.

1.8.1 Basis of Fig. 1–2 Giving Minimum Acceptance Numbers

In determining the function f_1 involved in equation (1–1), the average number of pieces inspected per lot, I, is treated as the dependent variable. Since a sample is always taken, n pieces will be inspected from every lot submitted. The number of times that the remainder of the lot $(N - n)$ will be inspected on the average is determined from the expression giving the probability that more than c defectives will be found in n. The sample is assumed to be drawn from a product of which a fraction, $\bar{p}$, is defective. The probability that c or less defectives will be found in n pieces selected at random from a product containing $\bar{p}$ fraction defective is given by the sum of the first $c + 1$ terms of the Point Binomial $[(1 - \bar{p}) + \bar{p}]^n$. Hence the average number of pieces inspected per lot is determined from the relation

$$I = n + (N - n) \left[1 - \sum_{m=0}^{m=c} C_m^n (1 - \bar{p})^{n-m} \bar{p}^m \right].$$

For the condition $\bar{p} < 0.10$, which is usual in practice, it has been found satisfactory to replace the Point Binomial by the Poisson exponential.* By multiplying both sides of the equation by p_t we obtain z in the form

$$z = M - (M - a) \sum_{m=0}^{m=c} \frac{(ka)^m e^{-ka}}{m!}, \quad (1\text{--}1a)$$

which is the function f_1 desired.

To obtain f_2, we state the probability of finding c or less defectives in a sample n taken from a lot N containing $M = p_t N$ defectives. This is given by the equation

$$P = \frac{1}{C_n^N} \sum_{m=0}^{m=c} C_{n-m}^{N-M} C_m^M.$$

But this equation is too difficult to handle in general computations on a large scale. When $p_t < 0.10$ and n is sufficiently large, a satisfactory approximation to the above equation may be developed from the first $c + 1$ terms of the Point Binomial; that is,

$$P = \sum_{m=0}^{m=c} C_m^n \left(1 - \frac{M}{N} \right)^{n-m} \left(\frac{M}{N} \right)^m, \quad \text{since} \quad p_t = \frac{M}{N}.$$

An even better approximation is obtained by interchanging † n and M in the latter equation, giving the expression

$$P = \sum_{m=0}^{m=c} C_m^M \left(1 - \frac{n}{N} \right)^{M-m} \left(\frac{n}{N} \right)^m.$$

Since $a/M \equiv n/N$, we obtain the final form

$$P = \sum_{m=0}^{m=c} C_m^M \left(1 - \frac{a}{M} \right)^{M-m} \left(\frac{a}{M} \right)^m, \quad (1\text{--}2a)$$

which is the function f_2 desired.

Now that we have f_1 and f_2 as expressed in equations (1–1a) and (1–2a) we must explain how Fig. 1–2 was obtained. When $P = 0.10$, for any pair (M, k), a particular pair (c, a) was found which made z a minimum. The acceptance number c may

---

* G. A. Campbell, "Probability Curves Showing Poisson's Exponential Summation," *The Bell System Technical Journal*, Vol. II, No. 1 (January 1923), pp. 95–113.

† Paul P. Coggins, "Some General Results of Elementary Sampling Theory for Engineering Use," *The Bell System Technical Journal*, Vol. VII, No. 1 (January 1928), p. 44, Equation (11).

assume only discrete values, since any piece must be considered either as defective or nondefective. Hence minimum values of z ($z_{\min}$) will be found for many pairs (M, k) for the same value of c. From this it is evident that on an M, k plane there exist zones in which the acceptance numbers are identical. To find the boundary lines of these zones it was noted that for certain pairs (M, k) two pairs of (c, a) exist, giving the same minimum value for z. These values of c were found to differ by 1 in all such cases. Designating in general two such adjacent acceptance numbers as c and $c + 1$ and corresponding values of a which satisfy the Consumer's Risk as a_c and a_{c+1}, we may obtain these boundary curves from the equation

$$(M - a_c) \sum_{m=0}^{m=c} \frac{(ka_c)^m e^{-ka_c}}{m!}$$
$$= (M - a_{c+1}) \sum_{m=0}^{m=c+1} \frac{(ka_{c+1})^m e^{-ka_{c+1}}}{m!}.$$

In using the above equation to determine these boundary curves for Fig. 1–2, the following steps were taken:

(1) Assume values for c and $c + 1$.

(2) Determine a_c and a_{c+1} for a given value of P, assuming N to be infinite.

(3) For any given value for k, solve the linear equation in M obtained by substituting the assumed values in the above equation.

(4) Using the value of M thus found, determine the exact values of a_c and a_{c+1} from equation (1–2a) for $P = 0.10$ (Fig. 1–3).

(5) Using the same value of k, again solve the linear equation in M, substituting the values of a_c and a_{c+1} obtained from step (4).

(6) If the values a_c and a_{c+1} obtained in step (4) satisfy the value of M thus found, the values of M and k define a point on the boundary curve between two adjacent acceptance numbers. If these values of a do not satisfy the value of M thus determined, steps (4) and (5) may be repeated until the limiting conditions are satisfied.

1.8.2 Basis of Fig. 1–3 Giving n for Any c

For given values of the Consumer's Risk P and acceptance number c, the sample size n may be obtained from equation (1–2a), since $a = p_t n$. For the case $P = 0.10$, values of a are presented in Fig. 1–3 for selected ranges of M and c.

1.8.3 Basis of Fig. 1–4 Giving the Minimum Average Number of Pieces Inspected per Lot

The curves in Fig. 1–4 represent specific values of z_{min} on an M, k plane for $P = 0.10$. Each curve was obtained by substituting given values of z_{min}, k, and (c, a) in equation (1–1a) and solving for M. If, for this value of M thus found, the selected values of c and a coincide with those read, respectively, from Figs. 1–2 and 1–3, a point was established for the given value of z_{min} on the M, k plane. If not, sufficient trials were made until the condition given by Figs. 1–2 and 1–3 were met. The curves for z_{min} were thus determined. To obtain I_{min} it was only necessary to use the relation

$$I_{min} = \frac{z_{min}}{p_t}.$$

Chapter 2

Single Sampling and Double Sampling
Inspection Tables

This chapter consists of material originally published in *The Bell System Technical Journal*, Vol. XX, No. 1 (January 1941), pages 1–61. In revising it for this edition, current standard terminology has been employed.

A CONSIDERABLE AMOUNT OF ATTENTION has been given to the application of statistical methods to problems of inspection, with emphasis on means for securing certain definite advantages, such as reduction in the cost of inspection, reduction in the cost of production by minimizing rejections, and the attainment of uniform quality of manufactured products.[1], [2], [3], [4] * This chapter describes four sets of sampling inspection tables (presented in Appendices 4, 5, 6, and 7) that have contributed in a notable way to important reductions in such costs and to substantial improvements in control of quality for many characteristics of products used in the Bell System.

Whether sampling may be employed to advantage in place of 100 per cent inspection usually depends, of course, on the purpose for which inspection is made. The sampling tables here presented provide definite procedures for conducting inspections that have certain immediate purposes which are described in some detail. Through their provision for instituting a "screening" inspection whenever quality falls below an acceptable level, the procedures have been found in practice to enforce a program of controlling quality in process as the alternative to high inspection costs.

2.1 GENERAL FIELD OF APPLICATION

The sampling tables presented in the Appendices have been developed for use in consumer or producer inspections of products composed of similar individual articles or pieces, in which it is desired to have assurance of a definite degree of conformance to specification requirements with a minimum of expense.

The following subsections indicate the general

* The references cited are listed at the end of this chapter.

conditions under which the tables are applicable, as well as some of the assumptions involved in their development.

2.1.1 Acceptance Inspection of Lots

The tables are intended for application in inspections whose immediate purpose is to determine the acceptability of individual lots of product.

By a lot will be meant a collection of individual pieces from a common source, possessing a common set of quality characteristics, and offered as a group for inspection and acceptance at one time. These pieces may be parts, partial assemblies, or finished units of product. For purposes of inspection, it is desirable that a lot be composed of pieces all of which have been produced under what are judged to be the same essential conditions. To this end, an attempt should be made to avoid grouping together batches of product that are likely to differ from one another in quality because of differences in the raw materials used or differences in manufacturing methods or conditions. For inspections made in a manufacturing plant, particularly where production is continuous, as with conveyor systems, the time element may often be the deciding factor in fixing the size of lot, and such items as convenience in handling and stocking or shipping facilities may make it desirable to take an hour's, a half-day's, or a day's production as the quantity to be considered as a lot for inspection purposes.

2.1.2 Quantity Production

Maximum advantage in the use of the tables may be expected for products produced more or less continuously on a quantity basis as distinguished from those produced intermittently on a small scale.

2.1.3 Inspection by "Method of Attributes"

Inspection by the "method of attributes" [5] is assumed. That is, each piece inspected is examined, gauged, or tested to determine whether it does or does not conform to the requirements imposed by specification.

For some characteristics, the requirements may be expressed as numerical limits to be met by the piece, such as maximum and minimum tolerance limits for a dimension or the minimum tolerance limit for the illumination of a lamp. For others, the requirements may be expressed in less precise terms, and inspection may consist in observing whether the piece does or does not conform to the finish, appearance, color, etc., of a standard sample or to the grade of workmanship commonly understood by the phrase "accepted standards of good workmanship."

2.1.4 Nondestructive Inspection

The tables are applicable primarily to quality characteristics that may be inspected by nondestructive means, so that at any time it is entirely practicable to inspect every piece in a lot. This is a consequence of the inspection procedure adopted in the development of the tables, wherein complete inspection of individual lots is prescribed under certain conditions.

However, the tables may also be used in selecting plans for destructive inspections—the OC curves (Operating Characteristic curves) of the plans are fully applicable but not their AOQL (Average Outgoing Quality Limit) values nor the minimum inspection feature. See the discussion in the Introduction, page 3.

2.1.5 Quality Measured by "Fraction Defective"

The yardstick of quality used in the tables is "fraction defective," that is, the ratio of the number of defective pieces to the total number of pieces under consideration.

A piece of product that fails to meet one or more of the requirements for the quality characteristics under consideration is referred to as *defective* and is called *a defective.** A *defect*, however,

---

* The original text is modified at this point and in subsequent paragraphs to use the term *defective*, instead of *defect*, and the term *acceptance number*, instead of *allowable defect number*, to provide consistency with Chapter 1 and current standard usage. (See, for example, ASQC Standard A1-1951, *Definitions and Symbols for Control Charts*, and ASQC Standard A2-1957, *Definitions and Symbols for Acceptance Sampling by Attributes*.)

is defined as a failure to meet a requirement for a single quality characteristic. It follows that when several characteristics are inspected an individual piece may have several *defects*, yet be only *a single defective*. For example, if, in the inspection of the "end illumination" of 1,000 lamps it were found that 10 of the lamps had illumination less than the minimum value specified and the remaining 990 had illumination equal to or greater than the minimum value, we would say that there were 10 defectives and that the lot of 1,000 was 1 per cent defective (fraction defective, $p = 0.01$).

2.1.6 Sampling Inspection

The tables are applicable where, under normal conditions, it will be satisfactory to inspect only a portion of the pieces in the lot and to accept the lot if the inspection results for this sample of pieces meet certain criteria. This, in effect, imposes the condition that it is not the purpose of this inspection to make sure that each piece in the lot conforms to the requirements for the characteristic inspected.

Such a situation is common, for example, in the process inspection of component parts of product units, where it may be the purpose of inspection to make reasonably certain that the quality passing on to the next stage is such that no extraordinary effort will be expended on defective parts. This situation is also common for various characteristics of finished units of product, such as some adjustment and dimensional items, items of condition, finish, and workmanship that can be covered by a "surface" inspection, as well as items for which 100 per cent inspections or tests have been made previously during process or are to be made in subsequent operations before delivery to the ultimate consumer. Characteristics whose conformance to specified requirements is of vital importance to the functional quality of the product, and for which 100 per cent inspection is feasible, may not, of course, be candidates for sampling inspection.

2.1.7 Acceptance Based on Observed Number of Defectives

The acceptance criterion used in the tables is a stated allowable number of defectives in a sample of stated size, referred to as the *Acceptance Number*.

If only one defective is allowed in a sample of n pieces selected from a lot, then the acceptance number is 1. The criterion for the acceptance of a lot is the finding of a number of defectives equal to or less than the acceptance number.

2.1.8 Random Samples

The theory used in the development of the tables assumes that each sample drawn from a lot is a random sample.

A random sample is one selected by a random operation,[6] such as would obtain if a number of physically similar chips, numbered to correspond to the pieces of product under consideration, were thoroughly mixed in a mixing bowl and a number of them, equal to the desired sample size, were withdrawn to identify which pieces of product should be included in the inspection sample.* When, in practice, there are indications that individual lots may be stratified in quality, it is, of course, best to select a "representative" sample, one such that each stratum or subportion of the lot is proportionately represented by a subsample that is selected by a random operation.

2.2 INSPECTION PROCEDURES

Two distinct methods of inspection are employed—single sampling and double sampling. In single sampling only one sample is permitted before a decision is reached regarding the disposition of the lot, and the acceptance criterion is expressed as an acceptance number, c. In double sampling a second sample is permitted if the first fails, and two acceptance numbers are used—the first, c_1, applying to the observed number of defectives for the first sample alone and the second, c_2, applying to the observed number of defectives for the first and second samples combined. The specific procedures assumed in the development of the tables are as follows:

Single Sampling Inspection Procedure

(a) Inspect a sample of n pieces.

(b) If the number of defectives found in the sample does not exceed c, the acceptance number, accept the lot.

(c) If the number of defectives found in the sample exceeds c, inspect all the pieces in the remainder of the lot.

(d) Correct or replace all defective pieces found.

Double Sampling Inspection Procedure

(a) Inspect a first sample of n_1 pieces.

(b) If the number of defectives found in the first sample does not exceed c_1, the accept-

* Use of one of the several published tables of random numbers is suggested.

ance number for the first sample, accept the lot.

(c) If the number of defectives found in the first sample exceeds c_2, the acceptance number for the combined first and second samples, inspect all the pieces in the remainder of the lot.

(d) If the number of defectives found in the first sample exceeds c_1 but does not exceed c_2, inspect a second sample of n_2 pieces.

(e) If the total number of defectives found in the first and second samples combined does not exceed c_2, accept the lot.

(f) If the total number of defectives found in the first and second samples combined exceeds c_2, inspect all the pieces in the remainder of the lot.

(g) Correct or replace all defective pieces found.

The double sampling procedure can, perhaps, be visualized more easily by reference to Fig. 2–1.

The theoretical development assumes that the inspection operation itself never overlooks a defect and that all defective pieces found, whether in samples or in the remainders of those lots that are inspected completely, will be corrected or replaced by conforming pieces.* Thus, lots that fail to be accepted by sample are assumed to be completely cleared of defects.

2.3 PROTECTION AND ECONOMY FEATURES

When a consumer † adopts sampling inspection in place of 100 per cent inspection, he forgoes the opportunity of assuring himself that each piece of product will conform to requirements and must choose a sampling plan that will provide a degree of protection against defective material that is consistent with his needs. This choice may be narrowed down by choosing some value of allowable

* While the mathematical solution assumes correction or replacement of defective pieces, it may be expedient practically to reject defective pieces and not replace them. The effect of following this, rather than the assumed procedure, involves differences in results too small to be of any practical consequence for the small values of per cent defective covered by the tables.

† The term "consumer" is used in the general sense of the recipient of the product after the inspection has been completed. This may, of course, be the ultimate consumer or his agent. However, in a manufacturing unit, if one department produces parts for use by a subsequent assembly department, the first department may be considered as the producer and the second the consumer.

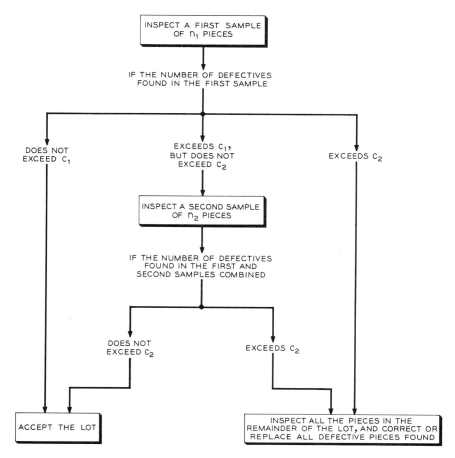

Fig. 2–1   **Double sampling inspection procedure**

per cent defective and by deciding whether this limiting value should apply to a limited quantity of product, such as a lot, or to the general output comprising a more or less steady flow of lots.

2.3.1 Two Kinds of Consumer Protection

For both the single sampling and double sampling procedures outlined above, tables are developed for each of the following two kinds of consumer protection:

(a) *Lot Quality Protection*—in which there is prescribed (1) some chosen value of limiting per cent defective in a lot (*Lot Tolerance Per Cent Defective*, LTPD) and also (2) some chosen value for the probability of accepting a submitted lot that has a per cent defective equal to the lot tolerance per cent defective. This probability is termed the *Consumer's Risk*.

(b) *Average Quality Protection*—in which there is prescribed some chosen value of *average* per cent defective in the product *after inspection* (*Average Outgoing Quality Limit*, AOQL) that shall not be exceeded no matter what may be the level of per cent defective in the product submitted to the inspector.

Single sampling plans employing the first of these two types of protection were developed in Chapter 1. An extension of the underlying theory as applied to double sampling will be given here. Sampling plans employing the second type of protection will likewise be covered for both the single sampling and double sampling procedures.*

---

* An adaptation of these concepts to inspection by the method of variables, using the arithmetic mean as an acceptance criterion, is given in a doctorate thesis (Columbia University) by H. G. Romig, "Allowable Average in Sampling Inspection," March 1939, for the case of a normally distributed characteristic that is statistically controlled with respect to the standard deviation.

The development of the second concept (AOQL) in 1927 was the result of a practical need in certain types of manufacturing process inspections, following considerable experience in the application of inspection procedures based on the first concept (LTPD and Consumer's Risk) which had been developed in 1924. Both have since been used extensively.

2.3.2 Minimum Amount of Inspection

For all of the four inspection plans covered, certain general principles, given in Chapter 1, are used.

For each plan two requirements are imposed— first, that the plan shall provide a specified degree of protection, as covered by (a) or (b) above, and second, that the amount of inspection shall be a minimum for product of *expected* quality, subject to the degree of protection imposed by the first requirement.

The first requirement can be satisfied by a large number of different combinations of sample sizes and acceptance numbers. The second requirement dictates which one of these combinations shall be chosen and requires a determination of the value of per cent defective to be normally expected in product submitted to the inspector. This expected value is referred to as the "process average" per cent defective.

For the inspection procedures here adopted, the amount of inspection that will be done *in the long run* is made up of two parts: (1) the number of pieces inspected in the samples and (2) the number of pieces inspected in the remainder of those lots that fail to be accepted by sample. We are to find a solution that will minimize the amount of inspection for uniform product * of process average quality.

In single sampling, for each combination of sample size and acceptance number, there will be a definite probability of exceeding the acceptance number for a sample drawn from uniform product of process average quality. This probability is termed the *Producer's Risk*. It represents the chance of not accepting a lot on the basis of the sample findings under these postulated conditions, and for the adopted inspection procedure it is thus the chance of inspecting the remainder of the

---

* By "uniform product" is meant one produced under conditions such that the probability of producing a defective piece remains constant at some definite value p. The solution thus provides for a minimum of inspection *if* quality is statistically controlled at a per cent defective level equal to the process average per cent defective.

pieces in the lot. The average (expected) amount of inspection per lot then equals the number inspected in the sampled portion plus the product of the Producer's Risk and the number of pieces in the remainder of the lot. This average value can be found for each combination, and the desired solution is obtained by choosing that combination of sample size and acceptance number for which the average amount of inspection is smallest.

In double sampling an entirely similar procedure is followed. Here, of course, we must consider the probability of taking a second sample when the first sample fails and then the probability of failure for the second sample. The over-all chance of failure constitutes the Producer's Risk for the complete double sampling plan.

No distinction is made at this point as to who actually inspects the remainders of those lots that fail to be accepted by sample. Whether the consumer does this inspection or rejects such lots and thus in effect requires the producer to do it will be considered immaterial for the purpose of solving the immediate problem. Interest will be centered only on the total amount of inspection done.

It should be noted that in the theoretical developments the number of defectives observed in a sample is not used to "estimate" the quality of the lot. Instead, it serves to indicate what action should be taken—whether the lot should be accepted, subjected to further sampling, or inspected completely—the entire process constituting a set of operations which when repeated over and over again produce a desired end result.

2.4 SINGLE SAMPLING— LOT QUALITY PROTECTION

The solution for this plan was given in Chapter 1, but it will be reviewed briefly, since certain of the principles and terms employed will be extended to the other three inspection plans.

Protection is defined by specifying values of

(a) Lot Tolerance Per Cent Defective (LTPD), the limiting per cent defective in a lot, and

(b) Consumer's Risk, the probability of accepting a lot of tolerance quality.

If the acceptance number is c, then the Consumer's Risk is the probability of finding c or less defectives in a random sample of n pieces drawn from a lot of N pieces in which the per cent defective is equal to the lot tolerance per cent defective. The tables are based on a Consumer's Risk of 0.10, a value found most useful in practice. For

Table 2–1

SOLUTION FOR A PARTICULAR CASE

Single Sampling, Lot Quality Protection

Sample Size n	Acceptance Number c	Probability of Acceptance by Sample	Probability of Inspecting Remainder of Lot (Producer's Risk)	Average Number of Pieces Inspected per Lot		
				In Sample	In Remainder of Lot	Total
75	0	0.713	0.287	75	265	340
125	1	0.891	0.109	125	95	220
170 †	2 †	0.958 †	0.042 †	170 †	35 †	205 †
210	3	0.984	0.016	210	13	223
250	4	0.994	0.006	250	5	255
290	5	0.998	0.002	290	1	291
325	6	0.999 +	0.000 +	325	0	325

n and *c* Combinations * — Application to Product Having Process Average Per Cent Defective = 0.45%

* These are n and c combinations for Lot Size, 1,000; Lot Tolerance Per Cent Defective, 3 per cent; and Consumer's Risk, 0.10.

† Plan involving minimum amount of inspection.

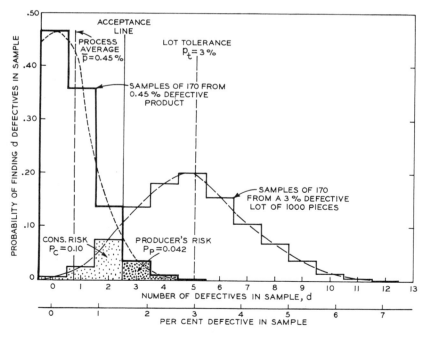

Fig. 2–2   Relation between Consumer's Risk and Producer's Risk

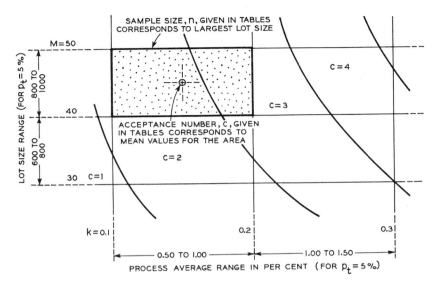

Fig. 2–3   Basis of choosing the n and c values given in the sampling tables

this choice, the chances of accepting a lot of worse than tolerance quality are less than 1 in 10.

For each value of c, such as 0, 1, 2, etc., there is a unique value of sample size, n, such that the probability of finding c or less defectives is 0.10. Any of these combinations of n and c will thus provide the desired consumer protection.

Now, for a given value of process average per cent defective, one of these combinations involves a smaller total amount of inspection than any of the others, as illustrated in Table 2–1. This combination of n and c, which provides the desired solution, gives the most efficient adjustment between Consumer's Risk and Producer's Risk from the standpoint of minimizing inspection effort. Figure 2–2 shows the relationship between these two risks for the conditions given in Table 2–1.

Curves providing a basis for solutions, such as that given in Table 2–1, appear in Chapter 1 for a Consumer's Risk of 0.10. The tables in Appendix 4 (Single Sampling, Lot Quality Protection) provide for practical use a complete set of such solutions for lot tolerance values from 0.5% to 10%. Each table is based on a particular value of lot tolerance per cent defective, and each solution, comprising a sample size, n, and acceptance number, c, covers a range of lot sizes and a range of process average values.* The value of n given in the tables is based on the largest lot size

for each lot size range, and the value of c corresponds to the mean lot size in each lot size range and to the mean value of process average in each process average range, as indicated in Fig. 2–3. This procedure is followed for all of the sampling tables presented in this book.

For the lot quality protection tables for both single and double sampling (Appendices 4 and 5), these choices are made to insure that, for the lot size range covered, the risk will not exceed the specified value (0.10) and to give *on the average*, for the process average range covered, the most economical plan. For reasons found advantageous in practice, sample sizes for samples of over 50 pieces are given to the nearest 5 units. For extremely large samples, the size is given to the nearest 10 units. This basis of rounding sample sizes is followed for all of the sampling tables presented in this volume.

On each table are listed values of AOQL to indicate the upper bound to the long-term average per cent defective in product after inspection that may be reached under the most adverse conditions.

2.5   DOUBLE SAMPLING— LOT QUALITY PROTECTION

The solution for this plan is carried out in substantially the same way as for single sampling. Protection is defined, as before, by specifying values of Lot Tolerance Per Cent Defective (LTPD) and Consumer's Risk. As for the single sampling procedure, a Consumer's Risk value of 0.10 is

---

* The extremely small process average range in the first column of each table has been specifically provided for those cases, increasingly common with long-continued use of these inspection procedures, where the process average per cent defective is for all practical purposes zero.

Table 2–2

COMPUTATION OF CONSUMER'S RISK

Double Sampling

Number of Defectives		Probability for $n_1 = 88$, $n_2 = 154$; Lot of 1,000 Pieces, 5% Defective
In the First Sample	In the Second Sample	
0		0.010 ⎤ Accepted by the
1		0.048 ⎦ first sample
2	0, 1, 2, 3, 4, or 5	0.018 ⎤
3	0, 1, 2, 3, or 4	0.015 ⎥
4	0, 1, 2, or 3	0.007 ⎥ Accepted by the
5	0, 1, or 2	0.002 ⎥ second sample
6	0 or 1	0.000 ⎥
7	0	0.000 ⎦
	Total Consumer's Risk	0.100

adopted. In double sampling a lot is given a second chance of acceptance if the first sample results are unfavorable, so that the Consumer's Risk is the sum of two parts: (a) the probability of accepting a lot of tolerance quality for the first sample and (b) the probability of its acceptance for the second sample if the first fails. For example, if the two acceptance numbers, c_1 and c_2, are 1 and 7, respectively, the Consumer's Risk is the sum of the probabilities for all of the different possible ways in which these criteria may be met, as shown in Table 2–2.

As in the case of single sampling, for any given process average value there is a large number of acceptance criteria—pairs of c_1 and c_2 in this case —for each of which sample sizes may be selected so as to give the desired Consumer's Risk of 0.10, but we wish to choose the combinations of n_1, n_2, c_1, and c_2 that will involve a minimum amount of inspection for product of process average quality. Furthermore, there is an unlimited number of ways of apportioning the Consumer's Risk between the first and second samples for each process average value. This latter factor introduces one more variable factor than will permit of a ready solution by other than trial and error methods, and accordingly an empirical choice has been made on the basis of a complete investigation of the relative practical advantages of several possible choices. Specifically, the solutions are based on an appor-

tionment such that the risk for the first sample is equal to the risk for an independent sample equal in size to the first and second samples combined. The use of an 0.06 risk in determining n_1 and $n_1 + n_2$ for given values of c_1 and c_2 provides a Consumer's Risk of almost exactly 0.10 over a considerable portion of the field covered by the tables, though in some areas a value as low as 0.056 is necessary. The "minimum" solutions for double sampling are, of course, conditioned by this choice.*

As shown in the mathematical supplement, Section 2.10, paired values of c_1 and c_2 that satisfy the condition of minimum inspection depend on (1) the tolerance number of defectives for a lot and (2) the ratio of the process average to the lot tolerance per cent defective. These values have been determined by trial and error and form the basis of the $c_1 c_2$ zones given in Fig. 2–7.

* Study of the effect of different apportionments of the Consumer's Risk on the average amount of inspection for product of process average quality indicates that considerably more than half of the 0.10 risk should be taken for small process average values and that less than half should be taken for large process average values. The single choice that was made provides a solution that closely approximates the true minimum over a large portion of the tables and was considered justified by the great saving in computation effort. With this choice, the average amount of inspection per lot does not in general exceed the true minimum by more than 3 to 5 per cent, although for extremely low process average values the excess may be as much as 15 per cent.

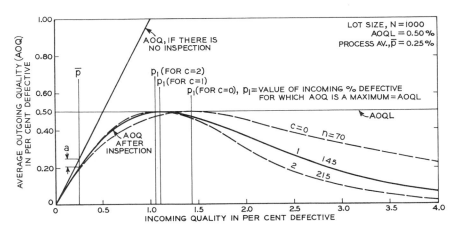

Fig. 2–4   Relationship between incoming quality, outgoing quality, and AOQL

The tables of Appendix 5 (Double Sampling, Lot Quality Protection) provide a complete set of solutions using paired values of c_1 and c_2 determined from Fig. 2–7. These tables are constructed on the same principles as the single sampling tables described in Section 2.4.

2.6 SINGLE SAMPLING— AVERAGE QUALITY PROTECTION

The solution for this plan considers the degree to which the entire inspection procedure screens out defectives in the product submitted to the inspector. Lots accepted by sample undergo a partial screening through the elimination of defectives found in samples. Lots that fail to be accepted by sample are completely cleared of defectives. The over-all result is some average per cent defective in the product as it leaves the inspector, termed the "average outgoing quality," which depends on the level of per cent defective for incoming product and the proportion of total defectives that are screened out.

The solid curve of Fig. 2–4 shows how the average outgoing quality varies for different values of incoming quality for a lot size of $N = 1,000$, a sample size of $n = 145$, and an acceptance number of $c = 1$. The curve is based on the concept of incoming product of uniform quality treated mathematically as an homogeneous universe. As the level of incoming per cent defective gets higher and higher, more and more lots are completely inspected. In turn, the average outgoing per cent defective increases, reaches a maximum value (0.50 per cent, in Fig. 2–4), and then falls off as a result of rapid increase in the amount of screening. This

maximum value is termed the Average Outgoing Quality Limit (AOQL).

For this plan, protection is defined by specifying a definite value of AOQL. For each possible value of c, such as 0, 1, 2, etc., there is a unique value of sample size that will give the specified value of AOQL. This is illustrated in Fig. 2–4. Any of these combinations of n and c provide the desired protection, and, as for the lot quality protection plans, we choose that combination of n and c that gives a minimum amount of inspection for uniform product of process average quality.

In the mathematical supplement, Section 2.10, it is shown that the acceptance number satisfying the condition of minimum inspection is dependent on two factors: (1) the number of defectives per lot for process average quality and (2) the ratio of the process average per cent defective to the AOQL value. Figure 2–9 defines zones of acceptance numbers for which the average amount of inspection is a minimum.

The tables of Appendix 6 (Single Sampling, Average Quality Protection) provide a complete set of minimum inspection solutions for AOQL values from 0.1 to 10 per cent. The choice of n and c for each solution in the tables is based on the procedure of Fig. 2–3 (using c zones given by Fig. 2–9) to ensure that the AOQL value over the area in question will not exceed the specified value and to give on the average for this area the most economical plan.*

On each table are given values of lot tolerance per cent defective for a Consumer's Risk of 10 per cent. These values are found useful in prac-

* OC curves for these single sampling plans are given in Appendix 1.

tice, since it is often desirable to know the degree of protection afforded to individual lots.

2.7 DOUBLE SAMPLING— AVERAGE QUALITY PROTECTION

The solution for double sampling differs from that for single sampling in that no simple relation has been found that gives directly the sample sizes that will result in a specified value of AOQL for a given lot size. This, together with the lack of simple relations for determining the choice of acceptance numbers (c_1 and c_2) that provide a minimum solution, has necessitated an empirical choice, the consequence of which is much the same as for the similar action taken in the solution of the problem of double sampling for lot quality protection.* Specifically, the interrelationship between n_1, n_2, c_1, and c_2 used in the latter case for a 10 per cent Consumer's Risk is used again here, and the solutions given are consequently minima that are contingent on this choice. An extensive trial-and-error investigation, using the underlying theoretical relations, leads to the conclusion that the degree to which the solutions given in these tables approach the true minima is of the same order of magnitude as for the double sampling tables for lot quality protection.

The method of solution is essentially that illustrated by example in Section 2.10.3. The pairs of values of c_1 and c_2 used in the solution are confined to those given in Fig. 2–7. For each of these pairs of c_1 and c_2, sample sizes are determined, using the above-mentioned relationship to a 10 per cent Consumer's Risk, that will give the desired AOQL value. Of these several sets of c_1, c_2, n_1, and n_2, that one is selected which involves the least amount of inspection.

The tables of Appendix 7 (Double Sampling, Average Quality Protection) provide a complete set of such minimum inspection solutions for AOQL values from 0.1 to 10 per cent. The choice of n_1, n_2, c_1, and c_2 for each solution in the tables is based on the general procedure of Fig. 2–3 (using the zones given in Fig. 2–7) to ensure that the AOQL value over the area in question will not exceed the specified value and to give on the average for this area the most economical plan.†

As for the single sampling AOQL tables, there are listed values of lot tolerance per cent defective for a Consumer's Risk of 10 per cent. In

* See the footnote on page 28.
† OC curves for these double sampling plans are given in Appendix 2.

this case these values have entered directly into the solution as just explained.

2.8 APPLICATION OF SAMPLING TABLES

In the above description of the sampling tables attention has been confined to the inspection of a single characteristic. The tables are, however, equally applicable to a group of characteristics considered collectively provided defects with respect to these characteristics are of essentially the same seriousness and may, therefore, be considered additive. When such application is made, the per cent defective values given in the tables embrace all such defects collectively, and, since more than one defect may occur on a single piece of product, any acceptance number listed in the tables should, by agreement, be considered either as a "number of defectives" or as a "number of defects." *

The sampling tables based on lot quality protection (the tables in Appendices 4 and 5) are perhaps best adapted to conditions where interest centers on each lot separately—for example, where the individual lot tends to retain its identity either from a shipment or a service standpoint. They have been found particularly useful in inspections made by the ultimate consumer or his purchasing agent for lots or shipments purchased more or less intermittently.

The sampling tables based on average quality protection (the tables in Appendices 6 and 7) are especially adapted for use where interest centers on the *average* quality of product after inspection rather than on the quality of each individual lot and where inspection is, therefore, intended to serve, if necessary, as a partial screen for defective pieces. The latter point of view has been found particularly helpful, for example, in consumer inspections of continuing purchases of large quantities of a product and in manufacturing process inspections of parts where the inspection lots tend to lose their identity by merger in a common storeroom from which quantities are withdrawn on order as needed.

Other things being equal, the average amount of inspection for double sampling is less than for single sampling. Figure 2–5 † gives a direct com-

* See the Introduction, foot of page 3.
† The curves and figures on this chart should be regarded as approximate. The mathematical relations involved are such that there exist unique values to be plotted on the M–k plane when certain approximate probability equations, referred to in Section 2.10, are employed in the solution, but not when exact equations are employed.

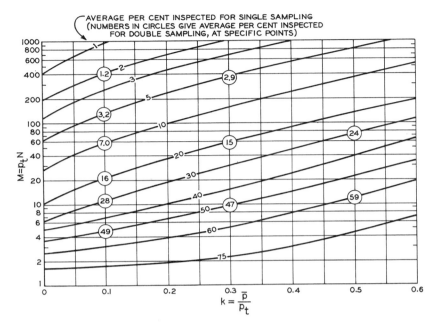

Fig. 2–5   Relative amount of inspection, double and single sampling

parison for the lot protection tables of Appendices 4 and 5. The saving obtained by using double instead of single sampling is greatest for large lot sizes and low process averages. Over the area of the tables found most useful in practice (percentage inspection less than 25 or 30 per cent), the saving generally exceeds 10 per cent and may be as great as 50 per cent. The saving that results from using the double sampling instead of the single sampling AOQL tables of Appendices 6 and 7 is of the same order of magnitude and may be estimated roughly from Fig. 2–5 by using the associated lot tolerance values listed in the AOQL tables for a chosen set of AOQL, lot size, and process average values. While the amount of inspection is a major cost item, other costs associated with double sampling frequently throw the advantage to single sampling. Among the added costs are those associated with interruption of work, extra handling of product, etc., incidental to the selection of an independent second sample. Aside from these considerations, it is common to find a psychological preference for double sampling. This appears to be associated with the tendency to look with favor on any plan that permits a "second chance" to make good, particularly when an initial failure is of a marginal character.

Given a specific problem of replacing a 100 per cent screening inspection by a sampling inspection, the first step is to decide on the type of protection desired, to select the desired limit of per cent defective—lot tolerance or AOQL value—for that type of protection, and to choose between single and double sampling. This results in the selection of one of the appended tables. The second step is to determine whether the quality of product is good enough to warrant the introduction of sampling. The economies of sampling will be realized, of course, only insofar as the per cent defective in submitted product is such that the acceptance criteria of the selected sampling plan will be met. A statistical analysis of past inspection results should first be made, therefore, in order to determine existing levels and fluctuations in the per cent defective for the characteristic or the group of characteristics under consideration. This provides information with respect to the degree of control of quality as well as the usual level of per cent defective to be expected under existing conditions. From this and other information is to be determined a value for the "process average" per cent defective that should be used in applying the selected sampling table, if sampling is to be introduced.

The determination of the process average per cent defective is an engineering problem, essentially one of prediction, in which use is made of all available information—knowledge of manufacturing conditions past and anticipated, judgment as to what periods of the past, if any, may be taken as representative of the future, results of analyses showing uniformity and level of per cent defective for such past periods, etc. The application of "control chart" analysis [1], [7] to past data is especially recommended.* If the process average value thus determined is within the range of process average values listed in the selected sampling table, then sampling can advantageously be introduced. If it is far outside this range, it would be quite satisfactory from a protection standpoint to use the last process average column of the selected table, but the sampling plan itself would force rejection or a screening inspection of such a large proportion of the lots that the introduction of sampling probably would not pay. If the process average value is but poorly estimated, the amount of inspection will be somewhat larger than need be, but the specified degree of protection will still be realized. Where there is uncertainty it is better to overestimate than to underestimate the process average value, since, for a given magnitude of error, a lesser amount of excess inspection will thereby be incurred.

It should be especially noted that the tables may be safely applied whether quality is well controlled or not. If, for example, the usual level of per cent defective is well within the range of process average values listed in the selected table but individual lots are frequently well outside this range, the sampling plan will usually permit acceptance by sampling while quality is good but force 100 per cent inspection when it is bad.

Experience with the tables indicates that where the procedures are used by a manufacturer within his own organization or by a consumer who rejects lots that are not accepted by sample, the general plan forces corrective action whenever quality becomes poorer than normally expected. The attendant increase in over-all inspection costs provides a compelling argument, in a language well understood by all, for determining the cause of trouble in the manufacturing process and for instituting measures for eliminating it as speedily as possible. Thus, while the inspection procedures have as their immediate purpose the provision of a curative technique whereby product already made is cleared of abnormal proportions of defects, they are found by experience to enforce the

adoption of a preventive technique—one that exerts economic pressure to track down and remove causes of abnormal quality variations, thus enforcing control of quality in the process and assuring better health in the product of tomorrow. Because of these factors, the long-term average outgoing per cent defective may rarely be expected to exceed one half the AOQL value associated with the inspection plan in use.

Quality control is achieved most efficiently, of course, not by the inspection operation itself but by getting at causes. [6] It may be expedited by carrying out regular statistical control analyses of the cumulative results of sampling inspection—preparing quality control charts [1], [7] for "per cent defective" with subgrouping of results on a lot-by-lot, a day-by-day, or a week-by-week basis—and making the findings available to those directly responsible for manufacturing processes.

Where a steady supply of product is offered for acceptance on a lot-by-lot basis, the use of these sampling procedures and tables, together with continuing control chart analyses of the inspection results obtained therefrom, have been found to provide a balanced and economical inspection program.

2.9 ACKNOWLEDGMENT

Work underlying the development and application of these tables has been contributed by many individuals in the Bell Telephone Laboratories and the Western Electric Company. The authors here express their indebtedness to these associates, particularly to those in the Western Electric Company who cooperated in the early development of the technical features of the plans and worked out shop procedures for use in their application. The laborious work of computing and preparing the tables in their final form was carried out by Miss Mary N. Torrey and Miss Ruth A. Bender; we wish to express our appreciation to them for their efforts to make the tables as free from error as possible.

---

* The following procedure has been useful. Tabulate the observed values of fraction defective, p, for at least 10 but preferably 25 immediately preceding lots (or groups of lots, say by days or weeks, if p is very small), excluding lots that are nonrepresentative for known reasons, and apply the control chart test to the observed values of p. If the data show statistical control, and if there are grounds for believing that future manufacturing conditions will be essentially the same as those of the past, use the average of the observed values of p as the process average value, $\bar{p}$. If lack

of statistical control is shown, replace values of p that are beyond $\pm 3\sigma$ control limits[1], [7] by values corresponding to $\pm 2\sigma$ control limits (where $\sigma = \sqrt{\bar{p}(1 - \bar{p})/n}$). Compute a corrected average value of p, in which the individually corrected values are used in place of the corresponding observed values. Unless other conflicting evidence predominates, use this corrected value as a tentative process average value until such time as a revision appears warranted on the basis of new evidence.

2.10   MATHEMATICAL RELATIONS

2.10.1   Fundamental Probability Formulas

The mathematical probabilities used in the solutions are based on equations corresponding to one or the other of the following two sets of conditions:

(a) Sampling from a finite universe.

(b) Sampling from an infinite universe.

In relations involving the determination of the Consumer's Risk the sample is considered as a sample from a lot of a finite number of pieces, and probabilities are correspondingly based on (a). For all other relations in the solutions—involving the determination of the Producer's Risk, the determination of the average number of pieces inspected per lot, etc.—the sample is considered as a sample from the general output of product—a source of supply—and probabilities are correspondingly based on (b).

Finite Universe

The probability of finding m defectives in a random sample of n units drawn from a finite universe (lot) of N pieces in which the number of defectives is $M = pN$ is given exactly by

$$P_{m,n,N,M} = \frac{1}{C_n^N} C_{n-m}^{N-M} C_m^M. \qquad (2\text{--}1)$$

When $p < 0.10$, a good approximation to (2–1) is given by the $m + 1$st term of the expansion of the binomial,

$$\left[\left(1 - \frac{n}{N} \right) + \frac{n}{N} \right]^M,$$

$$P_{m,n,N,M} \approx P_{m,\frac{n}{N},M} = C_n^M \left(1 - \frac{n}{N} \right)^{M-m} \left(\frac{n}{N} \right)^m. \qquad (2\text{-}1a)$$

When $p < 0.10$ and when $n/N < 0.10$, a good approximation to (2–1) is given by the $m + 1$st term of the Poisson exponential distribution,

$$P_{m,n,N,M} \approx P_{m,pn} = \frac{e^{-pn}(pn)^m}{m!}. \qquad (2\text{--}1b)$$

These are general equations applicable for any fraction defective, p, but are used in this chapter only for the specific case where $p = p_t$, the lot tolerance fraction defective, and where in turn, $M = p_t N$.

The Consumer's Risk, P_C, is the probability of meeting the acceptance criteria—c for single sam-

pling and c_1 and c_2 for double sampling—in samples drawn from a lot of N pieces containing exactly the tolerance number of defectives $M = p_t N$.

For single sampling,

$$P_C = \sum_{m=0}^{m=c} P_{m,n,N,M} \qquad \text{(when } p = p_t). \qquad (2\text{--}2)$$

For double sampling,

$$
\begin{aligned}
P_C = {} & \sum_{m=0}^{m=c_1} P_{m,n_1,N,M} \\
& + P_{c_1+1,n_1,N,M} \sum_{m=0}^{m=c_2-c_1-1} P_{m,n_2,N-n_1,M-c_1-1} \\
& + P_{c_1+2,n_1,N,M} \sum_{m=0}^{m=c_2-c_1-2} P_{m,n_2,N-n_1,M-c_1-2} \\
& + \cdots + P_{c_2,n_1,N,M} P_{0,n_2,N-n_1,M-c_2}
\end{aligned}
$$

$$\text{(when } p = p_t). \qquad (2\text{--}3)$$

Values of P_C in equations (2–2) and (2–3) are given approximately by substituting $P_{m,\frac{n}{N},M}$ or $P_{m,pn}$ for $P_{m,n,N,M}$ throughout, in accordance with equations (2–1a) and (2–1b), using $p = p_t$. The resulting equations will be referred to as (2–2a), (2–2b), (2–3a) and (2–3b), respectively.

$$\left. \begin{matrix} (2\text{--}2a) \\ (2\text{--}2b) \\ (2\text{--}3a) \\ (2\text{--}3b) \end{matrix} \right.$$

Infinite Universe

The probability of finding m defectives in a random sample of n pieces drawn from an infinite universe (general output of product) in which the fraction defective is p is given exactly by the $m + 1$st term of the expansion of the binomial, $[(1 - p) + p]^n$,

$$P_{m,n,p} = C_m^n (1 - p)^{n-m} p^m. \qquad (2\text{--}4)$$

When $p < 0.10$, a good approximation to (2–4) is given by the $m + 1$st term of the Poisson exponential distribution,

$$P_{m,n,p} \approx P_{m,pn} = \frac{e^{-pn}(pn)^m}{m!}. \qquad (2\text{--}4a)$$

The probability of meeting the acceptance criteria—c for single sampling and c_1 and c_2 for double sampling—in samples drawn from submitted product having a fraction defective of p is termed the probability of acceptance, P_a. For single sampling,

$$P_a = \sum_{m=0}^{m=c} P_{m,n,p}. \qquad (2\text{--}5)$$

For double sampling,

$$P_{a.} = \sum_{m=0}^{m=c_1} P_{m,n_1,p}$$
$$+ P_{c_1+1,n_1,p} \sum_{m=0}^{m=c_2-c_1-1} P_{m,n_2,p}$$
$$+ P_{c_1+2,n_1,p} \sum_{m=0}^{m=c_2-c_1-2} P_{m,n_2,p}$$
$$+ \cdots + P_{c_2,n_1,p}P_{0,n_2,p}. \quad (2-6)$$

Values of P_a in equations (2–5) and (2–6) are given approximately by substituting Poisson exponential probabilities, $P_{m,pn}$, for $P_{m,n,p}$ throughout in accordance with equation (2–4a). The resulting equations will be referred to as equations (2–5a) and (2–6a), respectively. $\quad$ (2–5a) (2–6a)

The Poisson exponential approximation is used in subsequent paragraphs wherever probabilities in sampling from an infinite universe apply. Tables [8] and charts [9], [10] are available from which these probability values (single term values or cumulative values for "c or less defectives") may be read directly.* Figure 2–6 gives a cumulative probability chart for the Poisson exponential distribution, which is widely useful in the solutions involved.

The Producer's Risk, P_P, is the probability of failing to meet the acceptance criteria in samples drawn from product of process average ($\bar{p}$) quality. Using $p = \bar{p}$ in equations (2–5) and (2–6),

$$P_P = 1 - P_a \quad \text{(when } p = \bar{p}\text{)}. \quad (2-7)$$

2.10.2   Lot Quality Protection

Single Sampling

Given: Lot size (N), lot tolerance fraction defective (p_t), Consumer's Risk ($P_C = 0.10$), process average fraction defective ($\bar{p}$).

To find: Values of n and c that will minimize $\bar{I}$, the average number of pieces inspected per lot for product of process average ($\bar{p}$) quality.

The average number of pieces inspected per lot (I) for product of p quality is given by

* In this work use was made of more complete tables that give cumulative probabilities for pn values up to 100; these tables are published in "Poisson's Exponential Binomial Limit" by E. C. Molina, Switching Theory Engineer, Bell Telephone Laboratories (New York: D. Van Nostrand Company, 1942).

$$I = n + (N - n)(1 - P_a), \quad (2-8)$$

where P_a is given by equation (2–5). Substituting the approximation of equation (2–5a) gives

$$I = n + (N - n)\left(1 - \sum_{m=0}^{m=c} P_{m,pn}\right). \quad (2-8a)$$

$\bar{I}$ is a specific value of I and is obtained from equation (2–8a) by using $p = \bar{p}$. The value of c that makes $\bar{I}$ a minimum may be read from the chart of Fig. 1–2 in Chapter 1, which uses coordinates of $M = p_t N$ and $k = \bar{p}/p_t$ and is based on $P_C = 0.10$. The corresponding sample size n may be read from Fig. 1–3 in Chapter 1 (based on equation (2–2a)), from Fig. 2–6 if appropriate, or by direct computation from equation (2–2), (2–2a), or (2–2b), using $P_C = 0.10$.

Double Sampling

Given: Lot size (N), lot tolerance fraction defective (p_t), Consumer's Risk ($P_C = 0.10$), process average fraction defective ($\bar{p}$).

To find: Values of n_1, n_2, c_1, c_2 that will minimize $\bar{I}$.

The average number of pieces inspected per lot (I) for product of p quality is given by

$$I = n_1 + n_2\left(1 - \sum_{m=0}^{m=c_1} P_{m,pn_1}\right) + (N - n_1 - n_2)(1 - P_a). \quad (2-9)$$

where P_a is determined from equation (2–6a).

$\bar{I}$ is a specific value of I and is obtained from equation (2–9) by using $p = \bar{p}$. As outlined on page 28, the pair of values of c_1 and c_2 that makes $\bar{I}$ a minimum is determined by trial and error, conditioned by the choice that the Consumer's Risk of 0.10 be divided between the first and second samples so that the "initial risk" for the first sample is 0.06. Figure 2–7 gives such pairs of c_1, c_2 values, corresponding to values $M = p_t N$ and $k = \bar{p}/p_t$.

For the selected apportionment of Consumer's Risk, the sample sizes n_1 and n_2 may be determined approximately from the following equations, which are based on equation (2–1a),

$$0.06 = \sum_{m=0}^{m=c_1} C_m^M \left(1 - \frac{n_1}{N}\right)^{M-m} \left(\frac{n_1}{N}\right)^m,$$
$$0.06 = \sum_{m=0}^{m=c_2} C_m^M \left(1 - \frac{n_1 + n_2}{N}\right)^{M-m} \left(\frac{n_1 + n_2}{N}\right)^m.$$
$$(2-10)$$

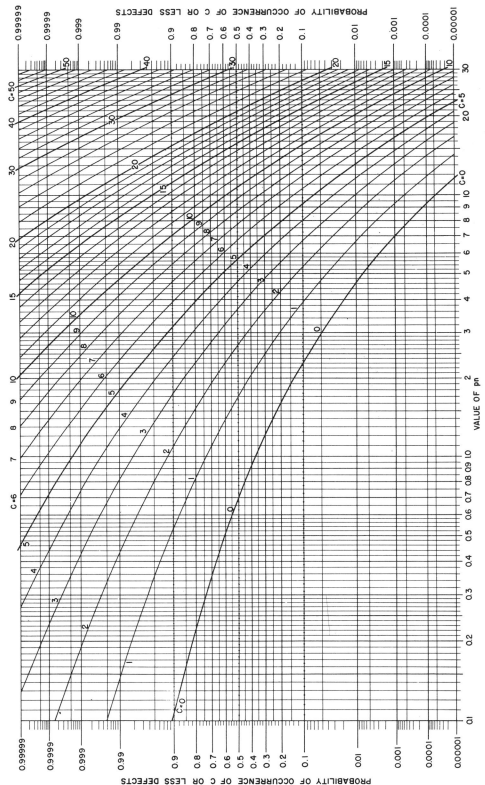

Fig. 2–6   Cumulative probability curves of the Poisson exponential distribution (a modification of a chart given by Miss F. Thorndike in *The Bell System Technical Journal*, October 1926). These curves may be used for determining the probability of occurrence of *c* or less defects in a sample of *n* pieces selected from an infinite universe in which the fraction defective is *p*. They may also be used as an approximation under certain conditions [Equations (2–1b) and (2–4a)] for determining the probability of occurrence of *c* or less *defectives* for a given *p* and *n*. Further, they serve as a generalized set of OC curves for single sampling plans, when the Poisson distribution is applicable.

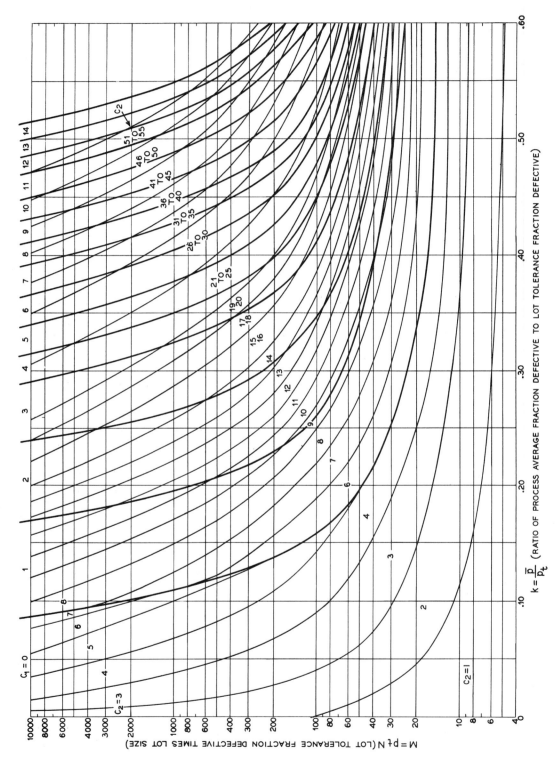

Fig. 2–7 Chart for determining acceptance numbers c_1 and c_2; lot tolerance protection, Consumer's Risk 0.10

Figure 2–8 based on these equations gives $p_t n_1$ and $p_t(n_1 + n_2)$ values associated with c_1 and c_2 for a given value of $M = p_t N$, and thus provides the desired values of n_1 and n_2.

The use of $P = 0.06$ for determining $n_1 + n_2$ corresponding to c_2 as well as for determining n_1 corresponding to c_1 results in a Consumer's Risk of approximately 0.10, as may be checked by writing the Consumer's Risk equation (2–3) as follows:

$$P_C = \sum_{m=0}^{m=c_1} P_{m,n_1,N,M}$$

$$+ \sum_{m=0}^{m=c_2} P_{m,n_1+n_2,N,M}$$

$$- \left(P_{0,n_1,N,M} \sum_{m=0}^{m=c_2} P_{m,n_2,N-n_1,M} \right.$$

$$+ P_{1,n_1,N,M} \sum_{m=0}^{m=c_2-1} P_{m,n_2,N-n_1,M-1} + \cdots$$

$$\left. + P_{c_1,n_1,N,M} \sum_{m=0}^{m=c_2-c_1} P_{m,n_2,N-n_1,M-c_1} \right).$$

$$(2\text{–}11)$$

The sum of the first two terms is 0.12, and the sum of the terms in parentheses is of the order of 0.02.

2.10.3 Average Quality Protection

General Relations

When the fraction defective in submitted product is p, the average quality after inspection (p_A) is given by

$$p_A = p \frac{N - I}{N} \qquad (2\text{–}12)$$

when all defective pieces found are replaced. If defective pieces found are removed but not replaced,

$$p_A = p \frac{N - I}{N - pI}, \qquad (2\text{–}12a)$$

the factor pI representing the average number of defective pieces removed. In deriving the tables, equation (2–12) has been used. The error in p_A resulting from the use of (2–12) rather than (2–12a) is pI/N, which is generally small.

The average outgoing quality limit (p_L) is the maximum value of p_A that will result under any sampling plan, considering all possible values of p in the submitted product. The value of p for

which this maximum value of p_A occurs is designated as p_1; hence

$$p_L = p_1 \frac{N - I}{N}. \qquad (2\text{–}13)$$

The value of p_1 for which $p_A = p_L$ may be determined by differentiating equation (2–12) with respect to p, equating to 0, and solving for p; i.e.,

$$\frac{dp_A}{dp} = \frac{N - I}{N} - \frac{p}{N} \frac{dI}{dp} = 0. \qquad (2\text{–}14)$$

Single Sampling

Given: Lot size (N), AOQL (p_L), process average fraction defective ($\bar{p}$).

To find: Values of n and c that will minimize $\bar{I}$.

The average quality after inspection (p_A), after substituting in equation (2–12) the value of I given in equation (2–8a), is obtained from the relation

$$p_A = p \frac{(N - n)}{N} \sum_{m=0}^{m=c} \frac{e^{-pn}(pn)^m}{m!}. \qquad (2\text{–}15)$$

Differentiating with respect to p in accordance with equation (2–14) gives

$$\frac{dp_A}{dp} = \frac{(N - n)}{N} \left[\sum_{m=0}^{m=c} \frac{e^{-pn}(pn)^m}{m!} - \frac{e^{-pn}(pn)^{c+1}}{c!} \right].$$

$$(2\text{–}16)$$

Equating to zero and solving for p gives the value of $p = p_1$ that makes p_A a maximum; i.e., $p_A = p_L$.

Let $p_1 n = x$; the particular case covered by equation (2–15) where $p = p_1$ and $p_A = p_L$ may then be expressed as

$$p_L = \frac{N - n}{Nn} x \sum_{m=0}^{m=c} \frac{e^{-x} x^m}{m!}, \qquad (2\text{–}17)$$

or

$$p_L = y \left(\frac{1}{n} - \frac{1}{N} \right), \qquad (2\text{–}18)$$

where

$$y = x \sum_{m=0}^{m=c} \frac{e^{-x} x^m}{m!}. \qquad (2\text{–}19)$$

Similarly, equation (2–16) equated to zero becomes, after substituting $p_1 n = x$ and simplifying,

$$\sum_{m=0}^{m=c} \frac{e^{-x} x^m}{m!} - \frac{e^{-x} x^{c+1}}{c!} = 0. \qquad (2\text{–}20)$$

Substituting the second term of equation (2–20) for the summation term in equation (2–19) gives

$$y = \frac{e^{-x} x^{c+2}}{c!}. \qquad (2\text{–}21)$$

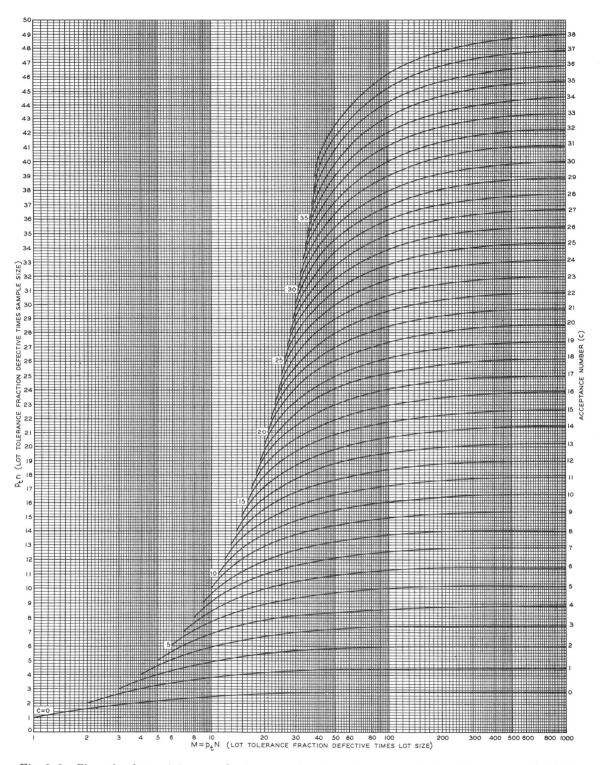

Fig. 2–8   Chart for determining sample sizes n_1 and n_2; lot tolerance protection, Consumer's Risk 0.10

These relations * provide a basis for determining the values of x and y, corresponding to specific values of c, listed in Table 2–3. The values of x were determined from equation (2–20) using Newton's Method of Approximation. The values of y were determined from equation (2–19).

The value of c that minimizes $\bar{I}$ [equation (2–8a), using $p = \bar{p}$], is given directly by Fig. 2–9, which uses coordinates of $\bar{M} = \bar{p}N$ and $\bar{k} = \bar{p}/p_L$. The curves bounding the c zones on Fig. 2–9 were obtained directly from relations between equations (2–18) and (2–8a), using $p = \bar{p}$, that define values of $\bar{M}$ and $\bar{k}$ such that $\bar{I}$ is the same for c and for $c + 1$.

The value of n corresponding to the value of c given on Fig. 2–9 is determined from equation (2–18), expressed as

$$n = \frac{yN}{p_L N + y}. \qquad (2\text{–}22)$$

* Reduction of the mathematical relations to this simplified form, and the determination of several x and y values, were contributed by the late Dr. Walter Bartky of the University of Chicago (when he was associated with the Western Electric Company) shortly after the development of the AOQL concept and the preparation of preliminary AOQL double sampling tables. Computation methods were contributed by Mr. George C. Campbell, formerly of the Bell Telephone Laboratories. A number of the values in Table 2–3 differ from those given in the first edition (more particularly the y values, but with relatively small differences for $c < 32$) as a result of recalculations suggested by Prof. Roger Lessard of the École Polytechnique, Montreal, P. Q., to whom we are indebted.

Example: Given: $N = 750$, $p_L = 0.01$, $\bar{p} = 0.004$.

To find: n and c.

Solution: $\bar{M} = \bar{p}N = (0.004)(750) = 3$; $\bar{k} = \dfrac{\bar{p}}{p_L} = \dfrac{0.004}{0.01} = 0.4$. Consulting Fig. 2–9, for $\bar{M} = 3$ and $\bar{k} = 0.4$, read $c = 1$. From Table 2–3, for $c = 1$, read $y = 0.8400$. From equation (2–22), $n = \dfrac{(0.8400)(750)}{(0.01)(750) + 0.8400} = 75.5$.

Sampling plan: $n = 76$, $c = 1$.

Double Sampling

Given: Lot size (N), AOQL (p_L), process average fraction defective $(\bar{p})$.

To find: Values of n_1, n_2, c_1, and c_2 that will minimize $\bar{I}$.

The average quality after inspection (p_A) is found by substituting in equation (2–12), the value of I given in equation (2–9).

$$p_A = \frac{p}{N}\left[(N - n_1)\sum_{m=0}^{m=c_1} P_{m,pn_1} + (N - n_1 - n_2) \right.$$
$$\times \left(P_{c_1+1,pn_1}\sum_{m=0}^{m=c_2-c_1-1} P_{m,pn_2} \right.$$
$$\left.\left. + \cdots + P_{c_2,pn_1}P_{0,pn_2} \right) \right]. \qquad (2\text{–}23)$$

Table 2–3

VALUES OF x AND y FOR GIVEN VALUES OF c

Used in equation (2–18) for determining p_L when N, n, and c are given, or in equation (2–22) for determining n when N, c, and p_L are given

Given c	x	y	Given c	x	y	Given c	x	y	Given c	x	y
0	1.00	0.3679	10	8.05	6.528	20	15.92	13.89	30	24.11	21.70
1	1.62	0.8400	11	8.82	7.233	21	16.73	14.66	31	24.95	22.50
2	2.27	1.371	12	9.59	7.948	22	17.54	15.43	32	25.78	23.30
3	2.95	1.942	13	10.37	8.670	23	18.35	16.20	33	26.62	24.10
4	3.64	2.544	14	11.15	9.398	24	19.17	16.98	34	27.45	24.90
5	4.35	3.168	15	11.93	10.13	25	19.99	17.76	35	28.29	25.71
6	5.07	3.812	16	12.72	10.88	26	20.81	18.54	36	29.13	26.52
7	5.80	4.472	17	13.52	11.62	27	21.63	19.33	37	29.97	27.33
8	6.55	5.146	18	14.31	12.37	28	22.46	20.12	38	30.82	28.14
9	7.30	5.831	19	15.12	13.13	29	23.29	20.91	39	31.66	28.96
10	8.05	6.528	20	15.92	13.89	30	24.11	21.70	40	32.51	29.77

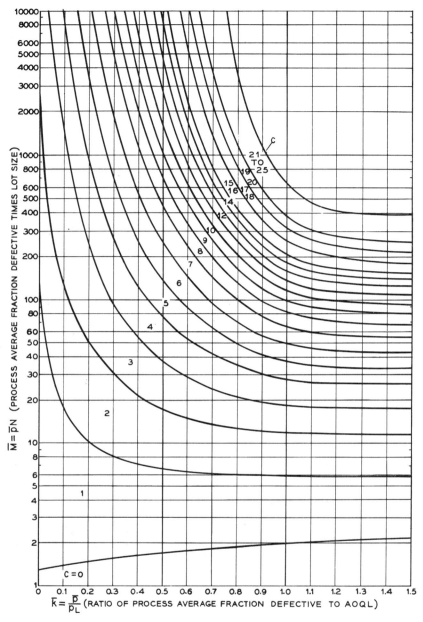

Fig. 2–9 Chart for determining the acceptance number, c; AOQL protection

Differentiating equation (2–23) with respect to p and equating to 0, in accordance with equation (2–14), and solving for p gives the value of $p = p_1$ that makes p_A a maximum; i.e., $p_A = p_L$. The resulting equation is not reproduced here, since it can be readily solved only for small values of c_1 and c_2. It is usually easier, particularly for the larger values of c_1 and c_2, to determine the maximum value of p_A (i.e., p_L) by trial and error, using work charts for estimating the region in which p_1 will be found.

The procedure used in preparing the tables and in finding the solution for a specific set of conditions is probably best illustrated by working out an actual example. In this procedure use is made of known relationships between p_t and p_L values, as given by the LTPD double sampling tables of Appendix 5, where an initial risk of 0.06 and a

Consumer's Risk of 0.10 are associated with p_t as outlined on page 28. For a given lot size, a work chart is prepared on which points corresponding to associated p_L and p_t values are plotted for pairs of c_1, c_2 values given in Fig. 2–7. A line drawn through all points for a single pair, such as $c_1 = 0$, $c_2 = 1$, indicates what p_t value should be associated with any p_L value specified. Figure 2–10 indicates the nature of the work chart and the following example illustrates its use.

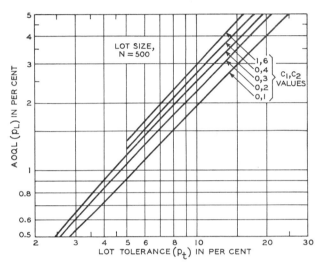

Fig. 2–10　Work chart giving p_t values corresponding to p_L values for given pairs of c_1, c_2 values; lot size $N = 500$

Example: Given: $N = 500$, $p_L = 0.01$, $\bar{p} = 0.004$.

To find: n_1, n_2, c_1, and c_2 that will minimize the average amount of inspection per lot. (Condition: For the associated lot tolerance value, p_t, the initial risk is 0.06 and the Consumer's Risk, $P_C = 0.10$).

Solution: Step 1—Consult the work chart, Fig. 2–10, for $N = 500$. Try $c_1 = 0$, $c_2 = 1$, and corresponding to $p_L = 0.01$, read $p_t = 0.054$.

Step 2—To determine whether the first choice of c_1, c_2 was the best.

$$M = p_t N = 0.054(500) = 27; \quad k = \frac{\bar{p}}{p_t}$$

$= \dfrac{0.004}{0.054} = 0.074$. Consult Fig. 2–7, giving the best c_1, c_2 values for given M and k values. Corresponding to $M = 27$, $k = 0.074$, read $c_1 = 0$, $c_2 = 2$. Hence the first choice was not the best.

Step 3--Similar to step 1. Consult the work chart, Fig. 2–10. For $c_1 = 0$, $c_2 = 2$, corresponding to $p_L = 0.01$, read $p_t = 0.047$.

Step 4—Similar to step 2 above. $M = p_t N = 0.047 (500) = 23.5; \; k = \dfrac{\bar{p}}{p_t} = 0.085$. Consult Fig. 2–7, and corresponding to $M = 23.5$, $k = 0.085$, read $c_1 = 0$, $c_2 = 2$. This agrees with the choice in step 3 and gives the desired solution.

Step 5—To determine n_1 and n_2 for $c_1 = 0$, $c_2 = 2$. On Fig. 2–8, corresponding to $M = 23.5$, for $c_1 = 0$, read $p_t n_1 = 2.67$; and for $c_2 = 2$, read $p_t (n_1 + n_2) = 5.60$. Since per step 3, $p_t = 0.047$, then $n_1 = 57$, $n_1 + n_2 = 119$, and $n_2 = 62$.

Sampling Plan. $n_1 = 57$, $n_2 = 62$, $c_1 = 0$, $c_2 = 2$.` (Rounding these values of n to the nearest 5 in accordance with the practice used in preparing the tables gives $n_1 = 55$, $n_1 + n_2 = 120$, $n_2 = 65$, the values shown in Appendix 7 in the table for AOQL = 1.0 per cent for $N = 401$–500, $\bar{p} = 0.21$ to 0.40 per cent.)

2.10.4　Nature and Magnitude of Errors

Each sampling plan (a combination of n and c values for single sampling and of n_1, n_2, c_1, and c_2 values for double sampling) in the tables constitutes a solution for a range of process average values and a range of lot sizes. The following paragraphs give information regarding the magnitude of errors associated with these solutions that may be present because of these two factors:

(1) Approximate equations and curves derived therefrom were used in place of exact equations over most areas of the tables in order to minimize computing effort.

(2) The sample sizes, n_1 and $n_1 + n_2$, listed in the tables, represent computed values rounded to the nearest unit for $n = 50$ or less, rounded to the nearest 5 for $50 < n < 1,000$, and rounded to the nearest 10 for $n > 1,000$.

Effect of Approximations

The percentage error in the Consumer's Risk value of 0.10, corresponding to lot tolerance values listed in the tables, attributable to the use of approximate equations and curves derived therefrom, is on the average about 3 per cent and should not exceed 7 per cent. The percentage error in the AOQL values listed in the tables, attributable to the use of approximate relations involving the Poisson exponential rather than the binomial distribution, is on the average about 4 per cent and should not exceed 12 per cent. In a large number of exploratory checks for both single and double sampling it was found in every instance that the Consumer's Risk and the AOQL values derived from approximate equations were larger than the corresponding exact values. The largest error observed in the Consumer's Risk for single sampling occurred when, instead of 0.10, the exact relation gave a value of 0.0937. Similarly the largest error in the AOQL occurred in single sampling when, instead of 0.0883, the exact relation gave a value of 0.0786. The observed errors in double sampling were of the same order of magnitude.

Effect of Rounding

The use of rounded values of n, n_1, and n_2 gives values of Consumer's Risk other than exactly 0.10. However, each sampling plan lists sample sizes based on the largest lot size in the corresponding lot size range. As a result, the Consumer's Risk associated with the p_t value designated at the top of the LTPD tables does not exceed 0.10, except in a few isolated cases where the risk may be as high as 0.12 for the largest lot size. Likewise, the AOQL value for any sampling plan in the AOQL tables does not exceed the value designated at the top of each table, except in a few isolated cases where the error due to rounding may be as much as 10 per cent of the designated value for the largest lot size.

The Consumer's Risk value of 0.10 and the AOQL values listed in the tables are therefore with few exceptions upper bounds that will not be exceeded in the application of the tables.

Nomenclature

N = number of pieces in the lot

n = number of pieces in the sample

n_1 = number of pieces in the first sample

n_2 = number of pieces in the second sample

c = acceptance number: the stated allowable number of defectives in a sample of stated size

c_1 = acceptance number for the first sample, n_1

c_2 = acceptance number for the first and second samples combined, $n_1 + n_2$

p_t = lot tolerance fraction defective; $100\ p_t$ = Lot Tolerance Per Cent Defective (LTPD)

p = fraction defective; also used specifically to denote the fraction defective in the submitted product

$\bar{p}$ = process average (expected) fraction defective in the submitted product

p_A = average fraction defective in the product after inspection: the Average Outgoing Quality (AOQ)

p_L = maximum value of the average fraction defective in the product after inspection: the Average Outgoing Quality Limit (AOQL)

p_1 = specific value of p in the submitted product, for which $p_A = p_L$

P_C = Consumer's Risk

P_a = probability of acceptance

P_P = Producer's Risk

I = average (expected) number of pieces inspected per lot for a submitted product of p quality

$\bar{I}$ = specific value of I when p in the submitted product = $\bar{p}$

$\bar{I}_{min}$ = minimum value of $\bar{I}$

M = $p_t N$ = number of defectives in a lot of tolerance (p_t) quality

$\bar{M}$ = $\bar{p} N$ = number of defectives in a lot of process average ($\bar{p}$) quality

k = $\bar{p}/p_t$ = ratio of process average fraction defective to tolerance fraction defective

$\bar{k}$ = $\bar{p}/p_L$ = ratio of process average fraction defective to AOQL

m = number of defectives found in the sample

C_n^N = $\dfrac{N!}{(N-n)!\,n!}$ = number of combinations of N things taken n at a time

References

1. W. A. Shewhart, *Economic Control of Quality of Manufactured Product* (New York: D. Van Nostrand Co., 1931).

2. E. S. Pearson, "The Application of Statistical Methods to Industrial Standardisation and Quality Control" (British Standards Institution, London, 1935).

3. H. F. Dodge and H. G. Romig, "A Method of Sampling Inspection," *The Bell System Technical Journal*, Vol. VIII, October 1929, pp. 613–631.

4. H. F. Dodge, "Acceptance-Rejection Requirements in Specifications," *Proc. A.S.T.M.* Vol. 34, Part II, 1934, pp. 877–890.

5. G. Udney Yule and M. G. Kendall, *An Introduction to the Theory of Statistics* (14th Ed.; New York: Hafner Publishing Co., 1950).

6. W. A. Shewhart, *Statistical Method from the Viewpoint of Quality Control* (The Graduate School, U. S. Dept. of Agriculture, Washington, 1939).

7. *A.S.T.M. Manual on Presentation of Data*, Supplement B (Philadelphia: A.S.T.M., 1940). *Note:* The revised edition is designated *A.S.T.M. Manual on Quality Control of Materials, 1951* (Part 3).

8. Karl Pearson, *Tables for Statisticians and Biometricians* (London: Cambridge University Press, 1914), Table LI.

9. G. A. Campbell, "Probability Curves Showing Poisson's Exponential Summation," *The Bell System Technical Journal*, Vol. II, January 1923, pp. 95–113.

10. Frances Thorndike, "Applications of Poisson's Probability Summation," *The Bell System Technical Journal*, Vol. V, October 1926, pp. 604–624.

Chapter 3

Using Double Sampling Inspection
in a Manufacturing Plant

This chapter consists of an article by D. B. Keeling and L. E. Cisne of the Western Electric Company which was first published in *The Bell System Technical Journal*, Vol. XXI, No. 3 (July 1942), pages 37–50. It is reprinted here by permission, with a few editorial notes and designations to adapt it to this Second Edition.

THE NECESSITY FOR QUALITY CONTROL in a manufacturing plant arises from the fact that all units of product cannot be made identical. To limit variations and attain controlled uniformity some sort of inspection must be established. It has been the experience of the Western Electric Company that quality control may be attained most economically by the use of a sampling inspection wherein only a portion of the entire output is examined for desired quality characteristics.

Advantages which have been gained through the use of sampling inspection, and with no adverse effect on previously existing quality levels, are a reduction in the cost of inspection by economies in inspection time; a reduction in the amount of scrap produced by making available for supervisory action useful records of the results of inspection; and, as an end result, the attainment of uniform quality of a satisfactory level.

It is the purpose of this chapter to provide a detailed method of procedure that has proved successful in establishing and maintaining one type of sampling—the "Average Outgoing Quality Limit" Double Sampling Plan. Statistically determined tables of lot sizes and corresponding sample sizes which guarantee a certain degree of protection have been used by the Western Electric Company for approximately fifteen years.[*] They were furnished by the Bell Telephone Laboratories and have recently been made generally available in an article published in the January 1941 issue of *The Bell System Technical Journal*.[†] A typical sampling table is shown in Fig. 3–1.

Briefly stated, the AOQL Double Sampling Plan involves the examination on a "go—no go" basis of a specified number of articles taken at random from a large group. The acceptance or rejection of this group is usually made on the basis of results obtained from the first sample alone. However, if the results from the first sample are not conclusive, an additional sample is examined before disposition of the lot is made.

The particular type of articles to which this plan has been applied are products consisting of individual parts, subassemblies, or completed apparatus, which, at the various stages of production where control is necessary, are the result of repetitive operations capable of considerable uniformity. The plan has also been applied to some extent on completed products and purchased materials where there is evidence that the product is of reasonable uniformity even though the quality history is meager or unavailable.

3.1 STEPS IN SETTING UP A DOUBLE SAMPLING LOT INSPECTION PLAN

3.1.1 Analysis of the Production Process

In order to determine the applicability of the Double Sampling Plan to existing inspection operations, it is necessary to examine the manufacturing and inspection processes and all data available, lot sizes, and process average values. The sampling table reproduced in Fig. 3–1 is based on an AOQL value of 1.5 per cent.

In tables prepared for shop use it has been found preferable to use a notation slightly different from that shown in Fig. 3–1; specifically to use AN instead of c to represent *Acceptance Number* and to use SS instead of n to represent *Sample Size*. The shop notation is used in the present chapter.

---

[*] That is, for approximately fifteen years prior to 1942, when this chapter was written. *Ed.*

[†] The AOQL Double Sampling Tables referred to are those in Appendix 7 of this edition. These tables give sample sizes and acceptance numbers for a variety of AOQL values,

Double Sampling Table for
Average Outgoing Quality Limit (AOQL) = 1.5%

Lot Size	Process Average 0 to 0.03%						Process Average 0.04 to 0.30%						Process Average 0.31 to 0.60%					
	Trial 1		Trial 2			p_t %	Trial 1		Trial 2			p_t %	Trial 1		Trial 2			p_t %
	n_1	c_1	n_2	n_1+n_2	c_2		n_1	c_1	n_2	n_1+n_2	c_2		n_1	c_1	n_2	n_1+n_2	c_2	
1–15	All	0	–	–	–	–	All	0	–	–	–	–	All	0	–	–	–	–
16–50	16	0	–	–	–	11.6	16	0	–	–	–	11.6	16	0	–	–	–	11.6
51–75	23	0	11	34	1	10.5	23	0	11	34	1	10.5	23	0	11	34	1	10.5
76–100	26	0	14	40	1	9.4	26	0	14	40	1	9.4	26	0	14	40	1	9.4
101–200	31	0	18	49	1	8.4	31	0	18	49	1	8.4	31	0	18	49	1	8.4
201–300	33	0	22	55	1	8.0	33	0	22	55	1	8.0	38	0	37	75	2	7.0
301–400	34	0	21	55	1	7.9	34	0	21	55	1	7.9	39	0	41	80	2	6.9
401–500	35	0	20	55	1	7.8	35	0	20	55	1	7.8	39	0	46	85	2	6.9
501–600	35	0	20	55	1	7.8	40	0	45	85	2	6.8	40	0	45	85	2	6.8
601–800	35	0	20	55	1	7.8	41	0	49	90	2	6.7	46	0	74	120	3	6.0
801–1000	36	0	19	55	1	7.8	42	0	48	90	2	6.5	47	0	78	125	3	5.9
1001–2000	44	0	51	95	2	6.3	44	0	51	95	2	6.3	49	0	81	130	3	5.7
2001–3000	45	0	50	95	2	6.2	45	0	50	95	2	6.2	55	0	110	165	4	5.3
3001–4000	45	0	50	95	2	6.2	50	0	85	135	3	5.5	55	0	115	170	4	5.2
4001–5000	45	0	50	95	2	6.2	50	0	85	135	3	5.5	55	0	120	175	4	5.1
5001–7000	46	0	54	100	2	6.1	50	0	90	140	3	5.4	60	0	155	215	5	4.7
7001–10,000	46	0	54	100	2	6.1	50	0	90	140	3	5.4	60	0	160	220	5	4.6
10,001–20,000	46	0	54	100	2	6.1	50	0	90	140	3	5.4	60	0	165	225	5	4.5
20,001–50,000	47	0	53	100	2	6.1	55	0	125	180	4	5.0	65	0	195	260	6	4.4
50,001–100,000	47	0	53	100	2	6.1	55	0	130	185	4	4.9	115	1	235	350	8	4.0

Fig. 3–1   Typical sampling table for the AOQL double sampling plan; AOQL = 1.5 per cent

able regarding past quality performance, such as records of per cent defective, consumers' complaints, etc. The following outline should serve not only as a measuring stick to determine the applicability of double sampling but also as an index of the conditions to be met for the successful use of double sampling with any inspection operation.

Composition of a Lot

The *lot*, or group of articles to be examined, should consist of product which is available in its entirety for acceptance or rejection at one time. For sampling purposes, the lot should have characteristics which are the result of a common system of causes. By this it is meant that the lot should, as far as possible, consist of articles made from relatively uniform raw material by operators of equivalent skill and by machines or methods of equivalent precision. If there is evidence of appreciable variation between corresponding machines, operators, or materials, it is desirable to confine a lot to the output of one machine, one operator, or one batch of material, in order to isolate a uniform group of product suitable for sampling inspection.

In brief, double sampling may be applied to the output of any repetitive unit operation capable of sufficient uniformity. However, unless immediately essential for economic reasons, it need not be applied at the particular time such an operation is completed, provided succeeding operations do not modify the inspection item under consideration.

Size of a Lot

In order to gain the maximum advantage from the use of double sampling, it is necessary that lots be as large as the limitation of uniformity will allow so that protection and control may be achieved with a minimum sample size relative to the number of units in the lot. The fact that proportionately smaller sample sizes are used with the larger lot sizes may be seen by reference to the Sampling Table of Fig. 3–1.

In sampling from larger lots it becomes *increasingly important* in practice to observe certain precautions in order to take care of instances where the lot may not be homogeneous; specifically, each sample should be a group of articles *taken at random from different locations throughout the lot* so that it will represent an impartial cross section of the lot.

Standard of Acceptability

It is necessary for the successful operation of any sampling inspection that at all times there be a known standard of acceptability for the individual article, that is, a definite description of the requirement for each inspection item, and reliable measuring equipment against which product may be conclusively checked. Practically, this condition will be realized when the characteristic is defined in such a manner that different observers will obtain consistent results.

Clearing Defectives

The theoretical background of the plan assumes the repair or elimination of all defectives * in samples of accepted lots, as well as all defectives in rejected lots before such lots are passed.

In view of this, close cooperation between production and inspection personnel is required in assuring that rejected lots are thoroughly inspected and cleared of all defectives found.

3.1.2 Selection of Proper Double Sampling Table

Tables are provided for a variety of Average Outgoing Quality Limit (AOQL) values and process average classes. The AOQL value is the maximum value of average per cent defective in the product after inspection which the sampling plan will assure over a long period of time, no matter how defective the product submitted for inspection may be. The process average is the normal per cent defective which is to be expected from the process.

To determine what AOQL value should be

---

* A *defective* is defined as an individual article that fails to meet the requirements for one or more inspection items. A *defect*, however, is defined as a failure to meet a requirement for a single quality characteristic for which inspection is made. It follows that when several characteristics are inspected an individual article may have several *defects*, yet be only a single *defective*.

adopted, it is necessary to decide upon a maximum average per cent defective which may be permitted in the product without serious consequences to the user. Product which is of such a nature that defects will be eliminated in subsequent operations may be assigned a rather generous AOQL, and, conversely, product which by its relation to the entire assembly may cause considerable inconvenience if it fails to meet requirements, usually warrants a strict AOQL. A rather generous AOQL may be assigned to inspection features which are considered relatively unimportant. In other cases, the use of sampling may be definitely inadvisable due to the importance of the requirement from a functional standpoint or from the standpoint of the possible effect of a failure upon the safety or health of an individual.

As a guide in the selection of the AOQL, the following table * is given, showing values that have been found to be satisfactory for the product listed. These percentage values of AOQL represent per cent of articles defective; if more than one defect is found on an article, the article is counted as one defective. Here, as elsewhere throughout this chapter, all figures relate to number of defectives and per cent of articles defective rather than number of defects.

The process average is commonly determined by summarizing the results of the first samples inspected during a representative period and may usually be obtained if there has been a previous inspection with associated records. In case suitable records are not available, an approximation may be made on the basis of an examination of a number of random samples selected from product of current manufacture. This will be a tentative figure and may require revision when data

---

* For further examples of choice of AOQL values, see the Introduction, p. 6. *Ed.*

Description	Requirements	AOQL
Machine screws	5 Dimensions	2.0%
Hexagon nuts	Visual inspection after zinc plating	2.0%
Twin eyelets	6 Dimensions and 4 visual requirements	3.0%
Relay coils †	Inductance and electrical breakdown	1.0%
Miscellaneous completed electrical apparatus	Resistance	0.5%

† This is a process check for these requirements which is supplemented by another sampling inspection after assembly.

MD-870-G (4-42)

LAYOUT FOR STATISTICAL SAMPLING INSPECTION LAYOUT NO. X

ISSUE NO. X DATE X

MATERIAL INSPECTED LAMPS CODE OR P.P. NO. X

MATERIAL FROM OPERATING DEPT. NO. XX PROCESS AVERAGE CLASS .31-.60% A.O.Q.L. 1.5 %

OPERATIONS DEPT. OPERATIONS Assembly

FOR USE BY INSP. SECT. X AUTHORIZED AND APPROVED BY X APPROVAL DATE X

LOT SIZE	SAMPLING SCHEME 1ST SAMPLE SS	AN	ADD	SS	AN
1-15	All	0	-	-	-
16-50	16	0	-	-	-
51-75	23	0	11	34	1
76-100	26	0	14	40	1
101-200	31	0	18	49	1
201-300	38	0	37	75	2
301-400	39	0	41	80	2
401-500	39	0	46	85	2
501-600	40	0	45	85	2
601-800	46	0	74	120	3
801-1000	47	0	78	125	3
1001-2000	49	0	81	130	3
2001-3000	55	0	110	165	4
3001-4000	55	0	115	170	4
4001-5000	55	0	120	175	4
5001-7000	60	0	155	215	5
7001-10000	60	0	160	220	5
10001-20000	60	0	165	225	5

INSPECTION OPERATIONS

Gage for over-all length and distance from end of terminal to end of bulb.

MATERIAL TO BE USED ON X

BREAKING POINT (ACTUAL) X %

EXPECTED QUALITY X % MAXIMUM AVERAGE % DEFECTIVES SATISFACTORY X

BREAKING POINT (ESTIMATED) X %

PROCESS AVERAGE ACTUAL RECORDS DATED

.557 %	%	%	%
FROM 7-6-40	FROM	FROM	FROM
TO 12-28-40	TO	TO	TO

PROCESS AVERAGE (ESTIMATED) X %

LAYOUT ISSUED BY X

APPROX. ANNUAL OUTPUT X USUAL LOT SIZE 4000 EST. HRS. PER YEAR 300

DELIVERY: CONTINUOUS [X] INTERMITTENT [ ] IRREGULAR [ ]

NON-ACCEPTABLE LOTS: DETAILED [ ] REJECTED [X] EITHER [ ]

OLD SAMPLING SCHEME:

USUAL LOT SIZE 2500 SAMPLE SIZE 100 A.O.Q.L. 1.5 % PROCESS AVERAGE .709 %

REMARKS:

Reissued to place in proper process average class.

NOTES BY STATISTICAL DEPT.

CHECKED BY: X

DATE X

Fig. 3–2 A typical layout for statistical sampling inspection

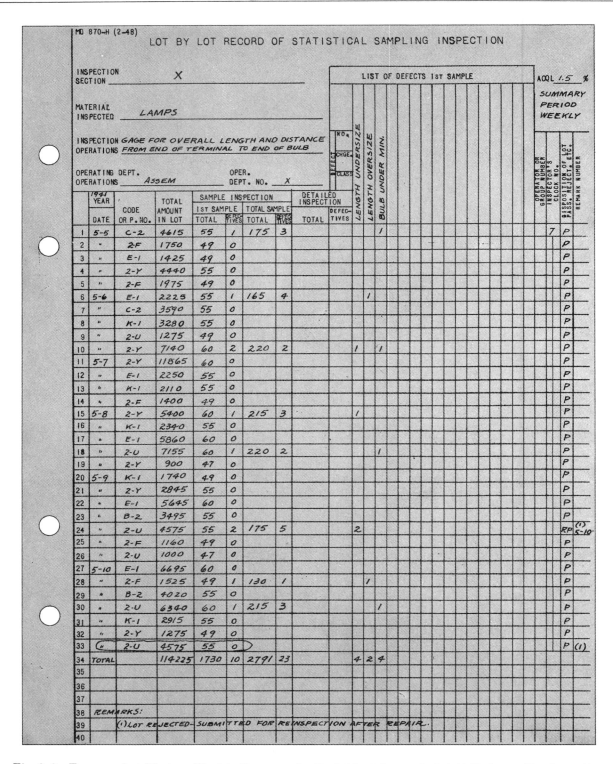

Fig. 3–3 Form used at Western Electric Company for the lot-by-lot record of statistical sampling inspection

accumulated from the operation of the Double Sampling Plan are available.

3.1.3 Issuance of the Inspection Layout—Fig. 3–2

After the AOQL value has been established, definite instructions in the form of a sampling layout (see Fig. 3–2) should be provided for the use of the inspector. The layout should contain a list of the inspection items for which inspection is required. It should also contain a copy of the sampling table selected, a description of the material to be inspected, the AOQL value, the process average, and other information which is of importance either in using or in maintaining the Double Sampling Plan.

3.1.4 Records

This section contains a description of the running records and clerical operations which are used for supervisory control. The records are "Lot by Lot Record of Statistical Sampling Inspection," (*sic*) and "Summary of Results of Inspection," shown in Figs. 3–3 and 3–4, respectively.

The Lot-by-Lot Record of Statistical Sampling Inspection—Fig. 3–3

On this form the inspector records the results of his inspection at the time of his observation. An examination of Fig. 3–3 will indicate how the form is filled out.

The Summary of Results of Inspection—Fig. 3–4

This form is kept with the Inspection Layout and the Lot-by-Lot Records. Entries are made as indicated on Fig. 3–4. Clerical operations involved in making the necessary computations are explained in the following paragraphs.

PA (Process Average)

The process average should be computed at least once every six months and more frequently when conditions warrant. Only data accumulated since the last computation should be used. When it is known that the quality of the product has changed significantly during the period, use only the data collected since the change.

Record the results of all process average computations and the periods covered by them in the space provided on the sampling layout.

Periodic Totaling of Lot-by-Lot Record, Fig. 3–3, and Posting on Summary of Results Form, Fig. 3–4

At suitable intervals the information on the Lot-by-Lot Record should be summarized to give the information below. Totals may, for contrast, be marked in red. Summaries after approximately twenty entries are generally considered satisfactory.

Number of lots
Number of articles in lots
Number of articles in first samples
Number of defectives in first samples
Number of articles in total samples
Number of defectives in total samples
Number of articles inspected during detailed inspection
Number of defectives found during detailed inspection

The above data are then posted on the Summary form, Fig. 3–4.

Per Cent Defective

The purpose of determining the per cent defective is to show the average quality of the product as received by the inspector during the period covered; it is obtained by dividing the total number of defectives found in first samples by the total number of articles inspected in first samples and multiplying by 100. Only the results of the first samples are used, in order to accord equitable treatment to all lots.

Control Chart

The per cent defective is plotted on the graph at the right of the Summary form (Fig. 3–4). Control limit lines are drawn around the process average to indicate the variation that may be expected due to random sampling. These control limits are determined by the following formulae: *

$$\left.\begin{array}{c}\text{Upper Control Limit}\\\text{for fraction defective}\end{array}\right\} = \bar{p} + 2\sqrt{\frac{\bar{p}(1 - \bar{p})}{n}}$$

$$\left.\begin{array}{c}\text{Lower Control Limit }\dagger\\\text{for fraction defective}\end{array}\right\} = \bar{p} - 2\sqrt{\frac{\bar{p}(1 - \bar{p})}{n}}$$

* The considerations involved in the establishment of control chart limits are discussed in ASA Standards Z1.1–1958 and Z1.2–1958, "Guide for Quality Control and Control Chart Method of Analyzing Data." In this case, 2-sigma limits have been chosen. For this particular application within the manufacturing plant over a period of years, this choice has appeared to strike an economic balance with respect to the net consequences of two kinds of "errors" that may occur in practice, namely: looking for trouble that does not exist and not looking for trouble that does exist. (*Note:* 3-sigma limits are now generally used in industry. *Ed.*)

† If this result is negative, the lower control limit is to be taken as zero.

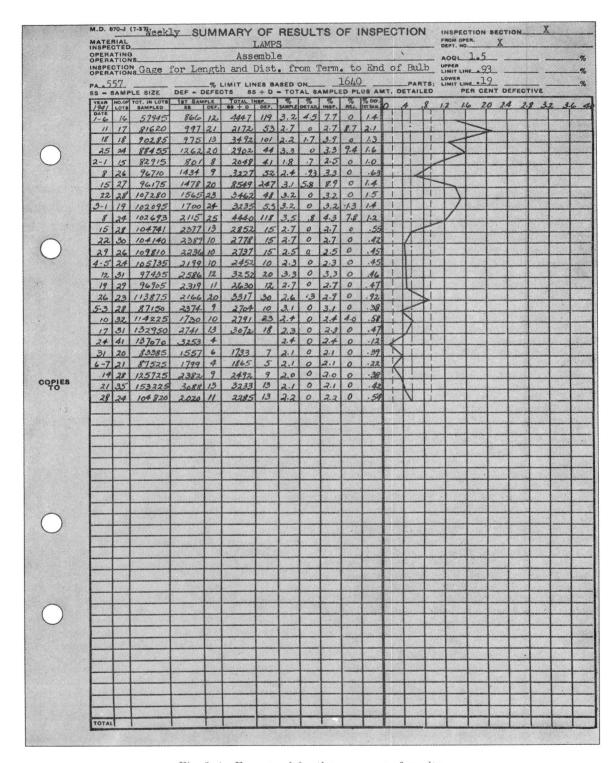

Fig. 3–4   Form used for the summary of results

where $\bar{p}$ = process average fraction defective and n = average number of articles in first samples inspected during the summary period.

For example, the records for a typical period of 10 weeks might show the following:

Total number of articles inspected in first samples = 16,400

Process average fraction defective = $\bar{p}$ = 0.00557 or 0.56 per cent

Thus, $n = \dfrac{16,400}{10} = 1640$ articles

Hence

Upper Control Limit

$$= 0.00557 + 2\sqrt{\frac{0.00557\,(1 - 0.00557)}{1640}}$$

$$= 0.00557 + 0.00368$$

$$= 0.00925 = 0.93\%$$

Lower Control Limit

$$= 0.00557 - 0.00368$$

$$= 0.00189 = 0.19\%$$

3.2 APPLICATION OF THE DOUBLE SAMPLING PLAN BY THE INSPECTOR

The chart of Fig. 3–5 illustrates in sequence the basic steps involved in the inspection of a lot.

An example of the operation of this chart is shown below:

Example:

AOQL   = 1.5%

Process Average = 0.56%

Lot Size   = 4615

On consulting the sampling table on the layout, which is information obtained from the table for AOQL = 1.5 per cent, Process Average Column

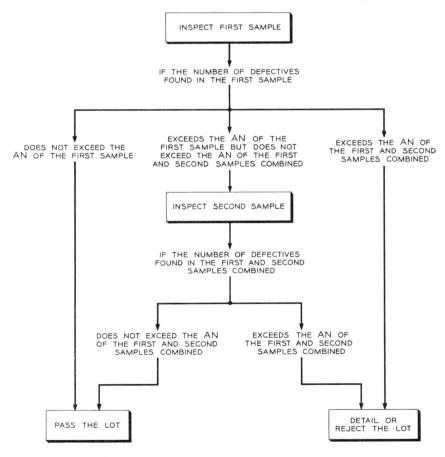

Fig. 3–5   Flow chart for double sampling

0.31 to 0.60 per cent (see Fig. 3–1), it will be found that for a lot of 4,001–5,000 parts the following sample sizes (SS) and acceptance numbers (AN) are shown:

Lot Size	First Sample			Total Sample	
	SS	AN	Add	SS	AN
4,001–5,000	55	0	120	175	4

This means that for the first sample 55 parts should be selected and completely inspected for all items covered by the sampling layout. In order that the per cent defective may be accurately determined for control purposes, a *complete inspection* of the first sample must be made regardless of how defective it may be. If five or more defectives are found in the sample of 55 parts, the lot should be rejected or detail inspected or otherwise disposed of as shown on the layout.

If no defectives are found in the 55 parts, the lot should be passed.

If one, two, three, or four defectives are found in the 55 parts, the second sample of 120 additional parts should be selected.

In the combined sample of 175 parts, a total of four defectives is allowed. If a total of five defectives is found before all of the 175 parts are inspected, sampling should be *discontinued* and the lot disposed of as indicated on the layout.

If less than five defectives are found in the combined sample of 175 parts, the lot should be passed.

3.2.1 Counting the Lot Size

The determination of the lot size may be made by weighing methods or careful estimates. It has been found in practice that an estimate which is within 20 per cent of the true value of lot size is satisfactory for sampling purposes.

3.2.2 Counting the Sample Size

Since the relation between corresponding lot sizes and sample sizes for a particular AOQL is not linear, the same order of accuracy of count does not prevail. If, in using a given sampling table, a sample smaller than that prescribed is taken, the result is to increase the AOQL; and if a sample larger than that prescribed is taken, the result is to decrease the AOQL and increase the cost of inspection.

Counting becomes a simple matter when articles are handled in compartment boxes or when the gauge or testing device provides for automatic count. Regardless of what method of counting is used, there should be agreement between the size of sample selected and sample size indicated in the sampling table.

3.2.3 Reinspection of Rejected Lots

In order to guarantee the protection promised by a particular Double Sampling Plan, it is required that a rejected lot be completely cleared of all defects. The rejected material should be repaired and returned for reinspection as one lot. It should then be reinspected for all inspection items, using the sample size that would normally apply to a new lot of the same size. If a defect is found during reinspection, it is evident that proper repair or sorting has not been accomplished. Such a lot should, of course, not be considered acceptable until all defects are removed.*

In the Lot by Lot Record the results of resampling of rejected lots are recorded on a separate line and the entry circled or otherwise identified so that it will not be included with the results of first samples on the Summary form.

3.3 SUPERVISION OF THE DOUBLE SAMPLING PLAN

In order that maximum advantage may be gained from the use of Double Sampling Lot Inspection, it is necessary that attention be given to the topics listed below. Upon proper attention to these factors depends the effectiveness of the plan.

3.3.1 Changes in AOQL

There is no assurance that the value originally selected for an AOQL will continue to be the most satisfactory in view of changing factors relating to the product, such as

(a) Changes in design of the product or changes in the requirements for inspection items may

---

Editorial Note: The inspection procedure to be followed for rejected lots that are corrected and resubmitted must, of course, be reasonable. If the sampling plan used for first submissions has an acceptance number, c, of 0 or 1, it is common practice to use the same n and $c = 0$ on resubmitted lots. If, however, $c > 1$ for first submissions, it may be found practicable, in some cases, to use $c = 1$ for resubmissions.

increase or decrease the trouble caused by the acceptance of defective parts and therefore will occasion a review of the AOQL value.

(b) New methods of manufacture that change the difficulty and therefore the cost of making a product will necessitate a reconsideration of the AOQL value.

(c) An excessive number of complaints or other reports of difficulty from succeeding stages of manufacture or from customers may indicate too large an AOQL.

(d) If no significant changes such as those mentioned above exist and there are repeated lot rejections, it would appear that quality is unsatisfactory. However, it should be determined whether or not succeeding operations can possibly tolerate more defectives than they are actually receiving and, if this is the case, the AOQL may profitably be increased.

3.3.2 Changes in Process Average

The economy of the sampling plan tends to decrease when the level of quality of the product shifts outside the range of process average on which the layout was based. The sampling table on the layout should ordinarily be changed whenever the process average shifts from one range to another. However, before reissuing the layout, the reason for the shift should be determined. If the process average has been reduced, an attempt should be made to make the change permanent; if it has been increased, the cause should be eliminated.

3.3.3 Interpretation of the Control Chart on the "Summary of Results of Inspection" Form

It is very important to review the control chart of per cent defective (Fig. 3–4) frequently, as it is an index of the success of the sampling plan. Not all fluctuations of the per cent defective are significant. Even though the quality of the product is controlled, the results of sampling inspection may produce fluctuations in the indicated per cent defective. These variations are measured by a simple control chart on the Summary form which employs control limits for values of per cent defective. As long as the plotted points of the per cent defective remain within control limits, the fluctuations are no greater than may reasonably

be expected from a uniform manufacturing process. However, if a point goes above the upper control limit, the cause may be defective workmanship, defective raw materials, or even a change in the severity of inspection. If a point goes below the lower control limit, the cause may be an improvement in quality or a change in the definition of a defect through misinterpretation or changes in inspection equipment or method of check. If the curve of the plotted points hugs either limit line or shows a definite trend toward one side or the other, a significant change in the quality of the product is indicated.

In order to achieve control of the quality of a manufactured product, direct and immediate action must be taken to stem unfavorable trends. The presence or absence of a satisfactory quality level may be detected by means of inspection, but such a level can only be originated and maintained by adequate manufacturing methods and equipment in the hands of a quality-minded producing personnel.

3.3.4 Changes in the Definition of a Defect

Either laxity or severity of inspection may cause the "reported" per cent defective to show significant changes, even though the actual per cent defective is unchanged. This may result from a change in the definition of a defect; that is, the same condition may at times be considered defective and at other times acceptable. This happens most frequently in border-line cases. To avoid such variations it is necessary that the condition that constitutes a defect be clearly defined and strictly followed in all inspections.

3.3.5 The Abnormal Existence of One Kind of Defect

When the sampling scheme includes the inspection for several different requirements, the acceptance number may at times be exceeded because of one kind of defect only. In other words, the lot would be satisfactory if this one kind of defect did not exist. If the same defect persists for several lots, it should receive definite supervisory attention. If substantial improvement is not feasible, it may be convenient to remove the inspection item to a separate sampling layout.

3.3.6 Abnormal Distribution of Defectives

Occasionally there may be reason to believe that a group of parts submitted for inspection is not uniform in quality throughout, that is, not a true lot as defined in the early part of the chapter.

Such a group should be divided into homogeneous sections and each section sampled separately. However, when this happens, subsequent lots for sampling purposes should be similarly subdivided, that is, they should be confined to the output of one source at a time; for example, one machine, one operator, or one batch of material, etc., based on one system of causes, so that control of quality at each source may be applied and the consumer protected from receiving spotty product.

3.4 IMPORTANCE OF CONSCIENTIOUS ADHERENCE TO PROCEDURE

In addition to the specific steps to be followed in establishing and operating a sampling plan, it must always be remembered that, since relatively important decisions concerning the acceptance of product hinge upon the results of an examination of a small group of parts, inspection must be conscientious and accurate. In order to guarantee the order of protection promised by the sampling plan, the results of inspection and the prescribed procedure for disposing of individual lots must be regarded with thorough respect.

Chapter 4

Operating Characteristics of Sampling Plans

This chapter, which was written especially for this Second Edition, describes the Operating Characteristic curves presented in Appendices 1, 2, and 3 and explains their relationship to the Sampling Inspection Tables of Appendices 4, 5, 6, and 7.

4.1 THE OPERATING CHARACTERISTIC CURVE

In the problem of choosing an acceptance sampling plan interest centers on how it will perform in actual practice, not merely for one value of fraction defective, p, in the material submitted for inspection, but for all possible values that might be encountered. This is perhaps best shown graphically by preparing what has been referred to as the "Operating Characteristic curve" * of the sampling plan, usually called its "OC curve," in which the probability of acceptance is plotted as a function of the fraction defective, p, of the material submitted for inspection.

There are two distinct types of OC curves, referred to here as Type A and Type B. These relate, respectively, to probabilities associated with two distinct sets of conditions referred to in Section 2.10, namely: (a) "sampling from a finite universe" and (b) "sampling from an infinite universe." These are discussed in Section 4.2.

The Appendices give a series of charts containing Type B OC curves for all the sampling plans in the AOQL Single Sampling and Double Sampling Tables of this book and for a general group of single sampling plans, arranged in three sets, as follows:

(1) *Appendix 1:* OC curves for the sampling plans in the AOQL Single Sampling Tables of Appendix 6.

(2) *Appendix 2:* OC curves for the sampling plans in the AOQL Double Sampling Tables of Appendix 7.

(3) *Appendix 3:* OC curves for a general set of single sampling plans for sample sizes

$n \leq 500$ and acceptance numbers $c = 0$, 1, 2, and 3.

The last of these three sets of curves provides a general reference set of Type B OC curves for single sampling plans based on binomial probabilities.

Type B OC curves for many of the Lot Tolerance Per Cent Defective (LTPD) sampling plans in Appendix 4 and Appendix 5 may be found as follows:

For LTPD Single Sampling Plans (Appendix 4)

(a) If $c \leq 3$ and $n \leq 500$, choose the proper curve in Appendix 3 or interpolate between two curves for the given c values, one for a slightly smaller value of n and one for a slightly larger value of n.

(b) If $c > 3$ or $n > 500$, choose a curve in Appendix 1 with the same c value and the closest value of n. (The lot size, N, is immaterial for Type B curves, as discussed later.) The tabulated AOQL value of the given sampling plan should be used as a guide in choosing the proper set of curves.

(c) As an alternative to (a) or (b) when conditions are such that the Poisson distribution can be considered as a satisfactory approximation to the binomial (see Sect. 2.10.1, p. 33), use the Poisson chart * of Fig. 2–6 directly to provide the desired OC curve.

---

* This term was originated by Col. H. H. Zornig shortly before World War II when he was Director of the Ballistic Research Laboratories at Aberdeen Proving Ground, Aberdeen, Md. The term used prior to that time was "probability of acceptance curve." (See the Introduction, p. 1.)

* The cumulative probability curves of the Poisson chart of Fig. 2–6 are in fact a set of generalized OC curves for single sampling plans, having a horizontal logarithmic scale for pn (p = fraction defective; n = sample size) and a vertical probability scale for P_a (probability of acceptance). For example, if an OC curve is desired for the plan: $n = 50$, $c = 3$, the pn scale is divided by 50 to give appropriate scale values of p (fraction defective) and values of P_a are read directly from the curve $c = 3$.

This chart gives substantially complete OC curves (Poisson) for values of c from 0 to 15.

For LTPD Double Sampling Plans (Appendix 5)

Using the tabulated AOQL value of the given sampling plan as a guide, choose the curve in Appendix 2 with the same value of c_1 and the nearest values of n_1 and c_2. The value of n_2 is less important than the values of n_1 and c_2 (and the lot size, N, is immaterial).

4.2 TWO TYPES OF OC CURVES

In the development of the sampling inspection tables two distinct sets of conditions, mentioned on page 33, were involved in defining the "risks" that are used. The specific concept used for the "Consumer's Risk" was associated with the notion of sampling from a *finite universe* or *lot* having a stated value of fraction defective, p. This particular risk is the probability of accepting a *lot* having a fraction defective, p, equal to the lot tolerance fraction defective, p_t. However, the specific concept used for the "Producer's Risk" (and employed in determining the AOQL of a sampling plan and in getting minimum-amount-of-inspection solutions) was associated with the notion of sampling from an *infinite universe* or *product* having a stated value of fraction defective, p. This particular risk is one minus the probability of acceptance of a lot coming from a *product* having a fraction defective, p, equal to the process average fraction defective, $\bar{p}$. The particular terms used in this discussion are thus defined:

Product: The conceptually infinite output of units * that the process would turn out if it continued in operation under the same essential conditions.

Product of Quality p (*Product Having Fraction Defective* p): The conceptually infinite output of units from a process for which the probability of producing a defective unit is p (which is equal to the fraction defective).

Product Quality p (*Product Fraction Defective* p): The fraction defective in product from a process for which the probability of producing a defective unit is p.

Thus, in sampling from a product, we consider what results are to be expected *in the long run*

* Variously referred to as units, pieces, parts, articles, etc.

for a series of lots considered as coming from the conceptually infinite output of units from a process. For this condition, the probabilities of acceptance are obtained by assuming a binomial distribution of lot quality values, which is just the distribution to be expected if product of quality p (the output of a process having probability p) were randomly subdivided into a collection of lots. It can be shown rigorously that random samples taken from a collection of lots having such a binomial distribution may be treated mathematically as if they were actually samples taken directly from an infinite universe having a fraction defective, p.

These two concepts, sampling from a *lot* and sampling from a *product* or *process*, provide the distinction between what have been called Type A and Type B operating characteristics. In turn, the values of probabilities and risks associated with Type A OC curves are called Type A probabilities and risks, and those associated with the Type B OC curves are called Type B probabilities and risks. Thus, while not so designated in the original published papers of Chapters 1 and 2, the Consumer's Risk, as used there, is, more specifically, the Consumer's A Risk, and the Producer's Risk, as used there, is the Producer's B Risk.

4.3 COMPARISON OF THE TWO TYPES OF OC CURVES

4.3.1 Type A OC Curves

The Type A OC curve of a sampling plan is defined as the curve showing the probability of accepting a lot as a function of *lot quality*, where lot quality is expressed as a fraction defective, p, or as a per cent defective, $100\,p$.

Consider the single sampling plan $n = 50$, $c = 1$, when applied to a lot of size $N = 500$.

The Type A OC curve of this plan when applied to $N = 500$ is shown by the heavy dashed line of Fig. 4–1 and is found by determining for various values of *lot* fraction defective, p, the probability of occurrence of 1 or less defectives in a random sample of $n = 50$ units drawn from a lot of $N = 500$ units containing pN defectives. Values were computed from the hypergeometric distribution corresponding to "sampling from a finite universe" (page 33). Thus the probability of acceptance is 0.95 for an 0.8 per cent defective lot of 500 units and 0.11 for a 7 per cent defective lot of 500 units.

In Fig. 4–1 are also shown the Type A OC curves

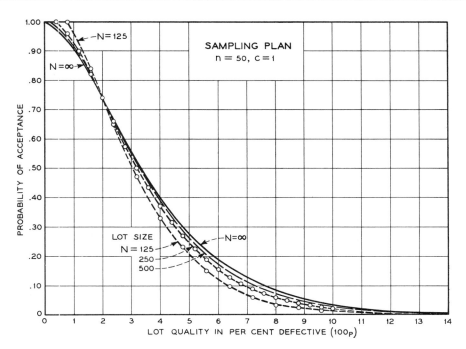

Fig. 4–1   Type A OC curves for various values of N, the lot size

of this plan ($n = 50$, $c = 1$) for several lot sizes in addition to $N = 500$, namely, $N = 125$, 250, and ∞.

The Type A OC curve for finite lot sizes is really a set of discrete points, since defectives can occur only as whole numbers. For example, for lots of $N = 125$, the only values of lot per cent defective that can exist are 0.0, 0.8, 1.6, 2.4, 3.2 per cent, etc., corresponding to the presence of exactly 0, 1, 2, 3, 4, etc., defectives in the lot. Such points are shown for the two smallest lot sizes in Fig. 4–1. For very large lot sizes, these points, of course, come very close together, giving a practically continuous curve, and for $N = \infty$ the curve is continuous.

For the sampling plan under discussion ($n = 50$, $c = 1$), a lot of size N containing exactly 1 defective ($p = 1/N$) would surely be accepted, which explains why, for smaller values of lot size N, the upper end of the Operating Characteristic curve intersects the P_a (probability of acceptance) $= 1.00$ line to the right of $100\,p = 0$. For every sampling plan having an acceptance number, c, equal to zero ($c = 0$), the OC curve stems from the point $P_a = 1.00$ and $100\,p = 0$.

From this group of curves, it will be noted that the lot size N has a relatively small effect on the characteristic curve so long as n/N is not large. The curves do not differ widely, even though the percentage sample size ($100 \times n/N$) increases up

to 20 per cent. It is emphasized that the *absolute* sample size, n, in number of units, is a much more controlling factor in determining the Type A OC curve of a sampling plan than the *percentage* sample size.

4.3.2   Type B OC Curves

The Type B OC curve of a sampling plan is defined as the curve showing the probability of accepting a lot as a function of *product quality*, where product quality is expressed as a fraction defective, p, or as a per cent defective, $100\,p$. The word "product" is used in the sense defined on page 56.

Consider the same single sampling plan as before: $n = 50$, $c = 1$.

The Type B OC curve of this plan is shown in Fig. 4–2 and is found by determining for various values of *product* fraction defective, p, the probability of occurrence of 1 or less defectives in a random sample of $n = 50$ units drawn from product of quality p. Values were computed from the binomial, corresponding to "sampling from an infinite universe" (pages 33–34).

Mathematically, the Type B OC curve of any sampling plan is identical with the Type A OC curve for $N = \infty$. (Compare the curve of Fig. 4–2 with the $N = \infty$ curve of Fig. 4–1.) Type B

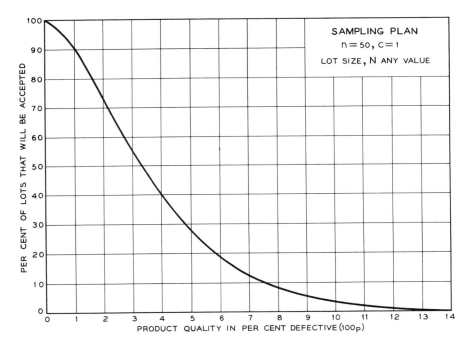

Fig. 4-2   Type B OC curve for lot size, *N*, of any value

curves are the most generally useful. They serve a dual purpose: they are correct Type B curves as such, but they may also be considered as suitable approximations to Type A curves, provided the sample size is small compared with the lot size (in general, if $n/N \leq 0.10$). Type B Operating Characteristic curves used for reference purposes are commonly provided with general captions, as indicated in the tabulation at the foot of this page, to permit interpretation on either a Type A or Type B basis, with the understanding, of course, that, for Type A use, they are limited in application to cases where n/N is small.

Figure 4-3 gives Type B OC curves for several different single sampling plans that have the same sample size but different acceptance numbers. This figure illustrates how drastically a change in acceptance number affects the probability of acceptance for any given value of product per cent defective.

Since the OC curves of Appendices 1, 2, and 3 are Type B curves, they show directly the "expected proportion of lots that will be accepted" for any assumed level of process quality. The Type B OC curve for a particular sampling plan in one of the AOQL sampling tables gives a quick answer to this type of question: "What percentage of the lots will be accepted (or rejected) if the process is running at a level of 2 per cent defective?" The curves can be used directly for finding what is re-

Description	Bottom Scale	Side Scale
General Caption	Fraction defective (or per cent defective)	Probability of acceptance
Meaning, for Type A	Lot quality, in fraction defective (or per cent defective)	Probability of accepting a lot of given quality
Meaning, for Type B	Product quality, in fraction defective (or per cent defective)	Probability of accepting a lot drawn at random from product of given quality. Also, expected proportion (or per cent) of lots that will be accepted, for product of given quality; or, proportion (or per cent) of lots expected to be accepted, for product of given quality.

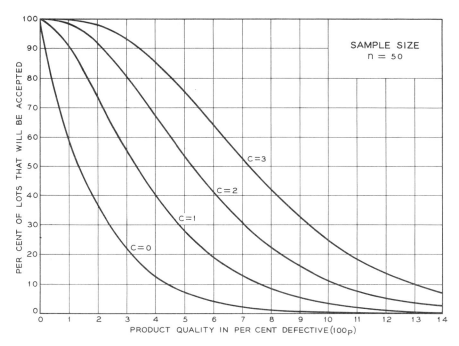

Fig. 4–3   Type B OC curves for single sampling plans with $c = 0, 1, 2,$ or 3

ferred to in Chapter 2 as the Producer's Risk of a sampling plan (the Producer's B Risk as defined on p. 56). This is found by reading the probability of acceptance value corresponding to the value of $p = \bar{p}$, the "process average per cent defective" (see Chapter 2), and subtracting it from unity.

These curves may be used as Type A OC curves for those plans in the AOQL sampling tables for which the sample sizes are small compared with the lot size. It is suggested that this use be made of the OC curves only when the sample size (in single sampling) and both the first sample size, n_1, and the second sample size, n_2, (in double sampling) of the sampling plan are less than 10 per cent of the lot size, conditions for which the binomial distribution is ordinarily considered a satisfactory approximation to the hypergeometric distribution. For example, when both n_1 and n_2 of a double sampling plan are less than 10 per cent of the lot size, these curves can be used for finding what is referred to as the Consumer's Risk in Chapters 1 and 2 and in the sampling tables (the Consumer's A Risk as defined on p. 56) —it is the probability of accepting a *lot* having a fraction defective, p, equal to the lot tolerance fraction defective, p_t. Under these conditions, the Consumer's Risk is found by reading directly on the OC curve the probability of acceptance value corresponding to $p = p_t$. When the sample sizes are a larger percentage of the lot size, the Type

A OC curve will fall somewhat below the Type B curve shown on the chart, as can be seen in Fig. 4–1 where the Type A OC curve for $N = \infty$ is identically the Type B OC curve.

4.4 COMPUTATION METHODS FOR OC CURVES

The exact formula, the sum of the first $(c + 1)$ terms of the expansion of the binomial $(q + p)^n$, was used for all OC curves for single sampling plans. This work was done quite recently with the aid of published tables.[1], [2], [3], [4] For areas not covered by these tables, values were obtained on an IBM 650 computer.

The computation work on the double sampling plans extended over a period of several years, most of it having been completed over five years ago, before binomial tables for $n > 150$ were available. Accordingly, for the larger sample sizes, Poisson tables [5] of probabilities were used, but it is believed that the errors of approximation are hardly appreciable in this graphical presentation. As a clue to areas of lesser accuracy, Fig. 4–4 shows the areas of the pn plane over which the Poisson gives two-or-more decimal accuracy. By two-or-more decimal accuracy is meant that for any single term of the binomial, the probability computed by the approximation formula does not differ from the true value for the binomial by more than

±0.005. It has been found that this accuracy is also obtained for cumulative probabilities of "*c* or less defects" for a majority of conditions involved in the AOQL sampling tables where the summation extends over a relatively small number of terms.

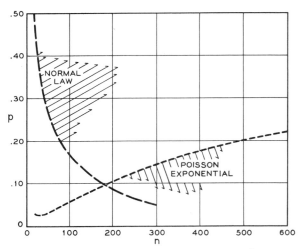

Fig. 4–4   **Chart of the *pn* area in which the Poisson exponential gives an accuracy within ±0.005 of the true value for the binomial**

As a general comment on the use of the Poisson as an approximation, we can say that for many engineering purposes the Poisson distribution can be considered as a satisfactory approximation *to the binomial distribution*, if *p* is less than 0.10, and *to the hypergeometric distribution*, if *p* is less than 0.10 and *n/N* is less than 0.10.

4.5   ACKNOWLEDGMENTS

The work of planning, computing, and charting the OC curves, extending over a period of years, was contributed by many individuals of the Bell Telephone Laboratories. The authors wish especially to express their appreciation to Mr. M. K. Kruger for organizing and directing the work, to Mrs. Elizabeth (Lockey) Breining, Miss Alice G. Loe, and Miss Judith Zagrodnick as principal long-term contributors to the computations and charting, to Miss Mary N. Torrey for technical assistance, and to the Misses Barbara Leetch and Melissa Twigg for work on the Single Sampling OC curves in the later stages of the project. We are also indebted to Mr. Spencer W. Roberts for IBM programming of computations for Single Sampling OC curves in areas not covered by published binomial tables.

References

1. "Tables of the Binomial Probability Distribution," U. S. Dept. of Commerce, National Bureau of Standards, Applied Math. Series 6; January 1950.

2. Harry G. Romig, *50–100 Binomial Tables*, John Wiley and Sons, Inc., New York, 1953.

3. "Tables of the Cumulative Binomial Probabilities," Ordnance Corps Pamphlet ORDP 20-1, Washington, 1952.

4. *Tables of the Cumulative Binomial Probability Distribution*, Harvard University Press, Cambridge, 1956.

5. E. C. Molina, *Poisson's Exponential Binomial Limit*, D. Van Nostrand Co., Inc., New York, 1942.

Appendix 1

OC Curves for

All Single Sampling Plans

in Appendix 6

Operating Characteristic Curves • Single Sampling Plans
Average Outgoing Quality Limit, AOQL = 0.1%

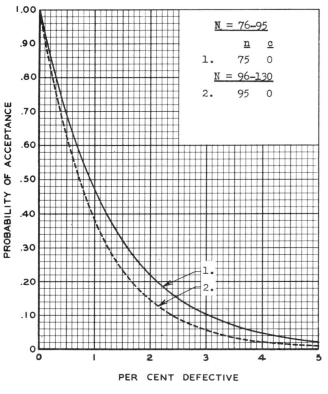

N = 76-95

	n	c
1.	75	0

N = 96-130

	n	c
2.	95	0

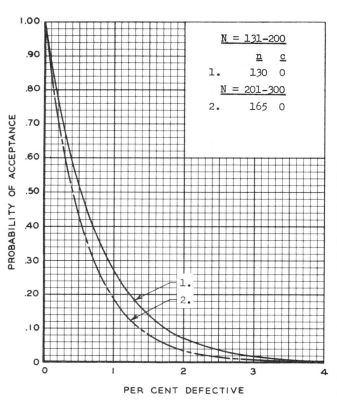

N = 131-200

	n	c
1.	130	0

N = 201-300

	n	c
2.	165	0

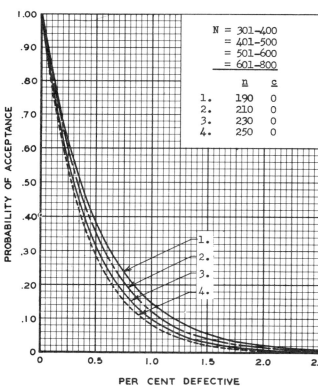

N = 301-400
= 401-500
= 501-600
= 601-800

	n	c
1.	190	0
2.	210	0
3.	230	0
4.	250	0

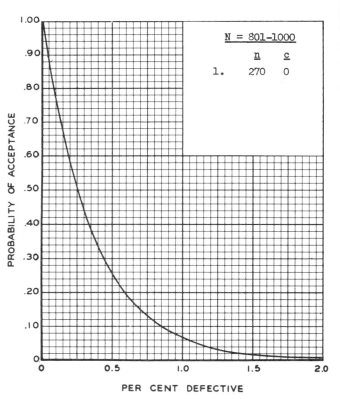

N = 801-1000

	n	c
1.	270	0

Operating Characteristic Curves • Single Sampling Plans
Average Outgoing Quality Limit, AOQL = 0.1%

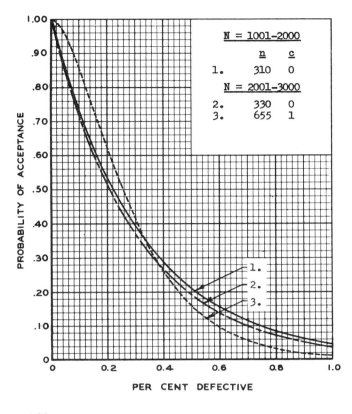

N = 1001-2000

	n	c
1.	310	0

N = 2001-3000

	n	c
2.	330	0
3.	655	1

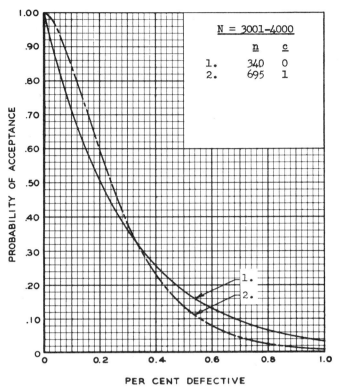

N = 3001-4000

	n	c
1.	340	0
2.	695	1

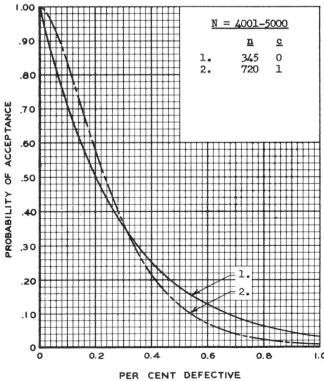

N = 4001-5000

	n	c
1.	345	0
2.	720	1

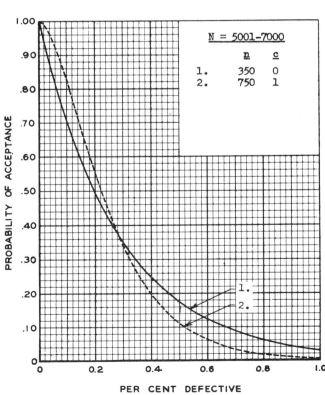

N = 5001-7000

	n	c
1.	350	0
2.	750	1

Operating Characteristic Curves • Single Sampling Plans
Average Outgoing Quality Limit, AOQL = 0.1%

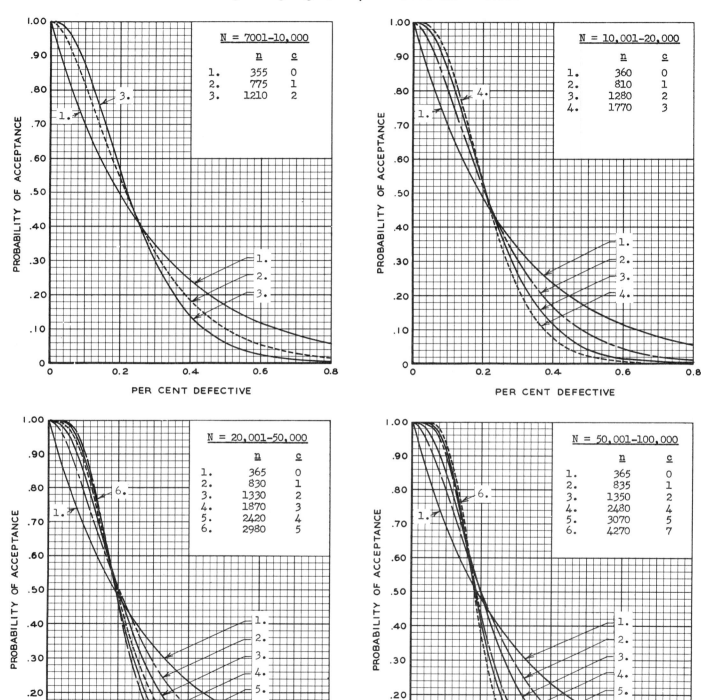

N = 7001-10,000

	n	c
1.	355	0
2.	775	1
3.	1210	2

N = 10,001-20,000

	n	c
1.	360	0
2.	810	1
3.	1280	2
4.	1770	3

N = 20,001-50,000

	n	c
1.	365	0
2.	830	1
3.	1330	2
4.	1870	3
5.	2420	4
6.	2980	5

N = 50,001-100,000

	n	c
1.	365	0
2.	835	1
3.	1350	2
4.	2480	4
5.	3070	5
6.	4270	7

Operating Characteristic Curves • Single Sampling Plans

Average Outgoing Quality Limit, AOQL = 0.25%

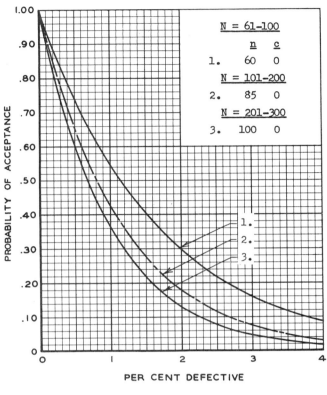

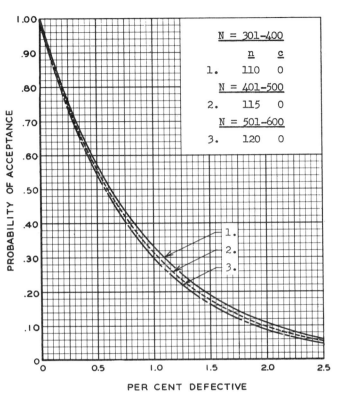

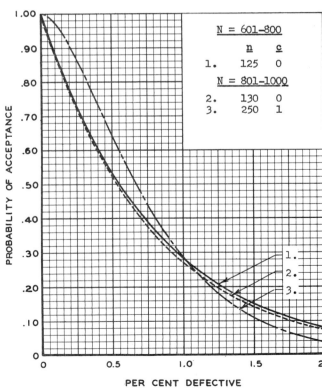

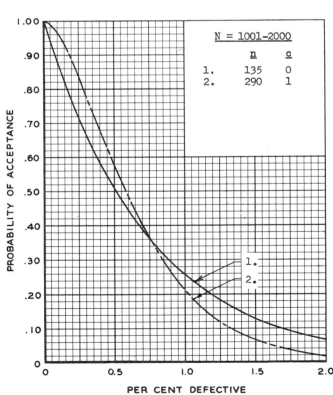

0.25%
AOQL

Operating Characteristic Curves • Single Sampling Plans

Average Outgoing Quality Limit, AOQL = 0.25%

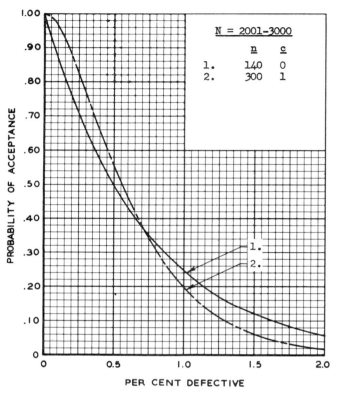

N = 2001-3000

	n	c
1.	140	0
2.	300	1

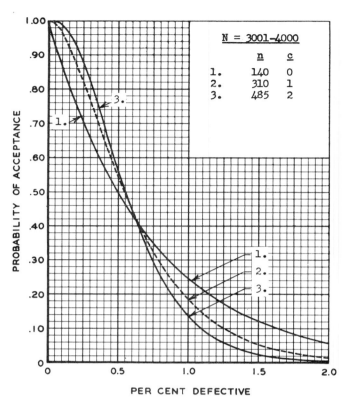

N = 3001-4000

	n	c
1.	140	0
2.	310	1
3.	485	2

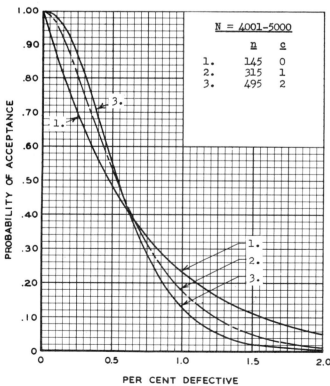

N = 4001-5000

	n	c
1.	145	0
2.	315	1
3.	495	2

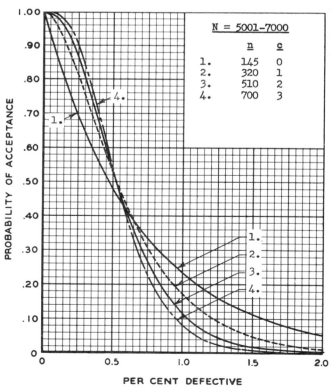

N = 5001-7000

	n	c
1.	145	0
2.	320	1
3.	510	2
4.	700	3

SINGLE
SAMPLING

0.25%

AOQL

Operating Characteristic Curves • Single Sampling Plans

Average Outgoing Quality Limit, AOQL = 0.25%

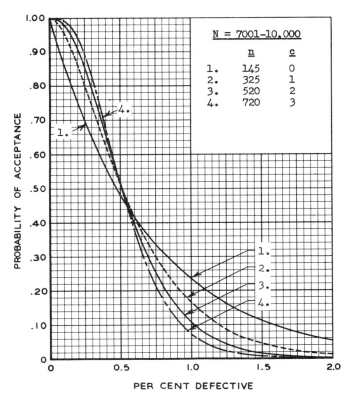

N = 7001–10,000

	n	c
1.	145	0
2.	325	1
3.	520	2
4.	720	3

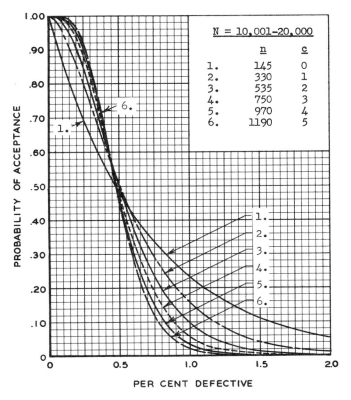

N = 10,001–20,000

	n	c
1.	145	0
2.	330	1
3.	535	2
4.	750	3
5.	970	4
6.	1190	5

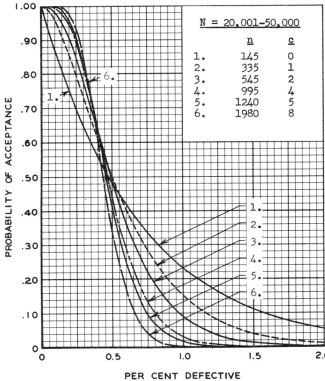

N = 20,001–50,000

	n	c
1.	145	0
2.	335	1
3.	545	2
4.	995	4
5.	1240	5
6.	1980	8

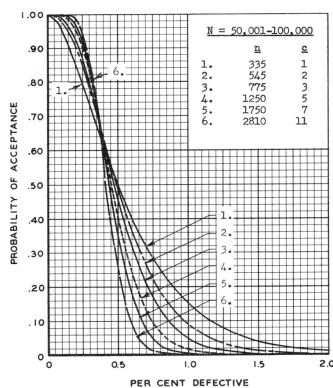

N = 50,001–100,000

	n	c
1.	335	1
2.	545	2
3.	775	3
4.	1250	5
5.	1750	7
6.	2810	11

SINGLE
SAMPLING
0.5%
AOQL

Operating Characteristic Curves • Single Sampling Plans
Average Outgoing Quality Limit, AOQL = 0.5%

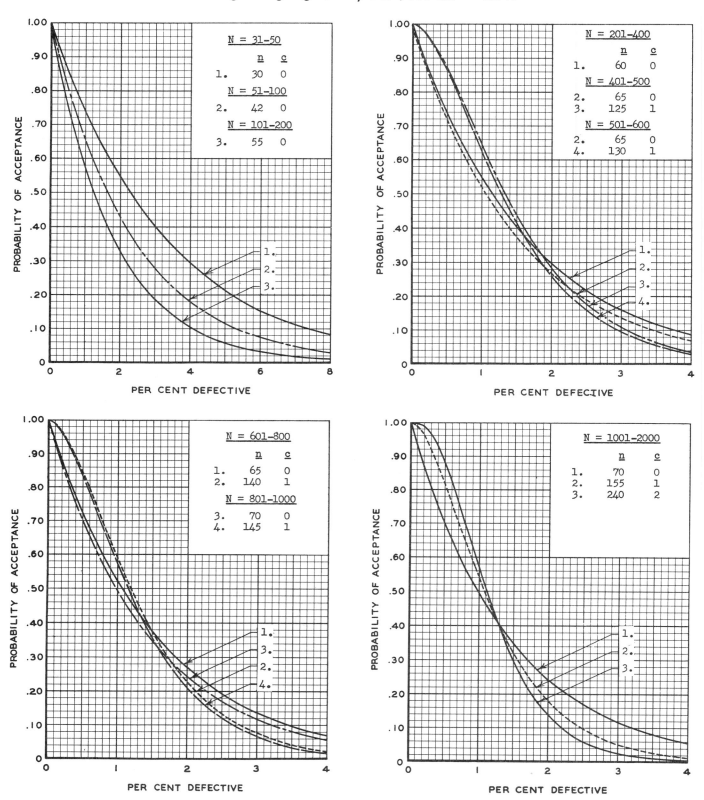

Operating Characteristic Curves • Single Sampling Plans
Average Outgoing Quality Limit, AOQL = 0.5%

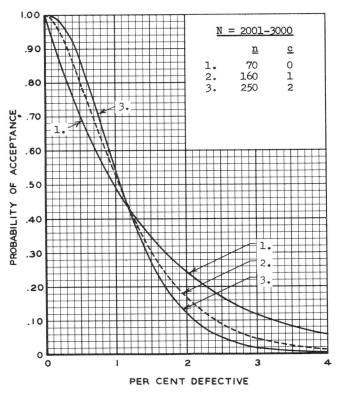

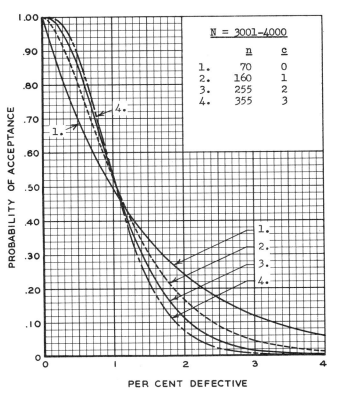

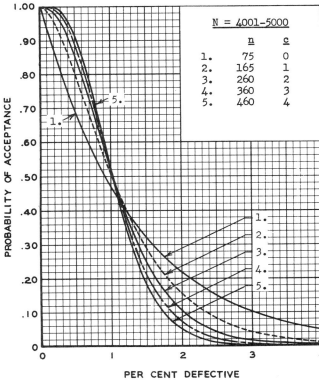

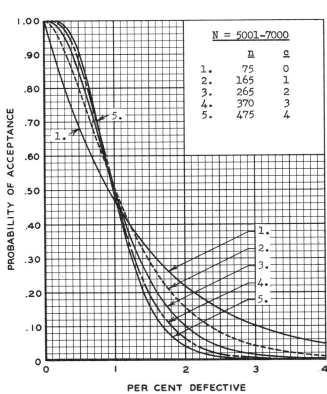

SINGLE
SAMPLING
0.5%
AOQL

Operating Characteristic Curves • Single Sampling Plans
Average Outgoing Quality Limit, AOQL = 0.5%

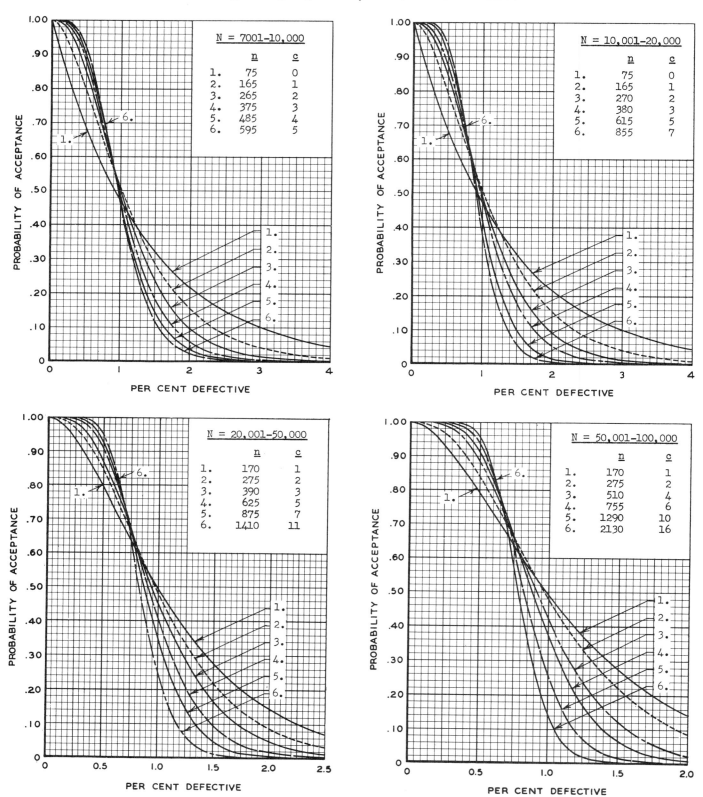

N = 7001–10,000

	n	c
1.	75	0
2.	165	1
3.	265	2
4.	375	3
5.	485	4
6.	595	5

N = 10,001–20,000

	n	c
1.	75	0
2.	165	1
3.	270	2
4.	380	3
5.	615	5
6.	855	7

N = 20,001–50,000

	n	c
1.	170	1
2.	275	2
3.	390	3
4.	625	5
5.	875	7
6.	1410	11

N = 50,001–100,000

	n	c
1.	170	1
2.	275	2
3.	510	4
4.	755	6
5.	1290	10
6.	2130	16

Operating Characteristic Curves • Single Sampling Plans

Average Outgoing Quality Limit, AOQL = 0.75%

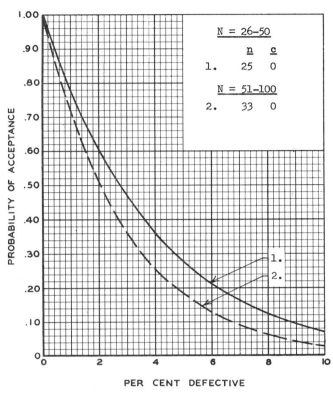

N = 26-50

	n	c
1.	25	0

N = 51-100

	n	c
2.	33	0

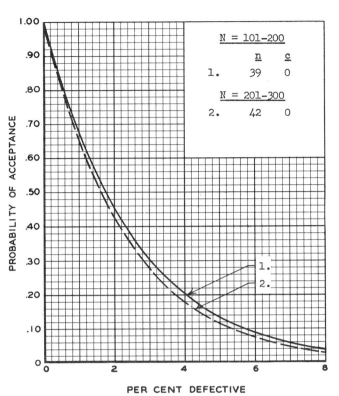

N = 101-200

	n	c
1.	39	0

N = 201-300

	n	c
2.	42	0

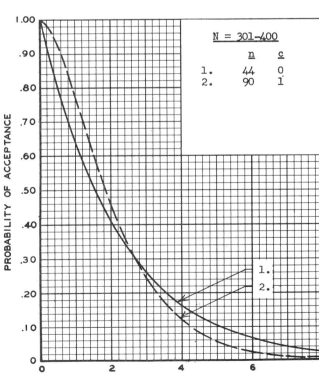

N = 301-400

	n	c
1.	44	0
2.	90	1

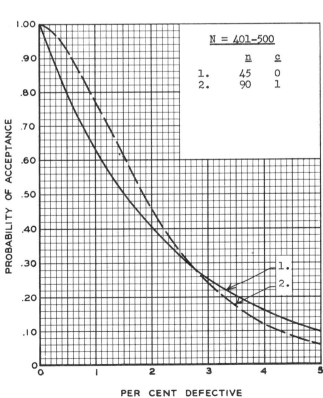

N = 401-500

	n	c
1.	45	0
2.	90	1

SINGLE
SAMPLING

0.75%

AOQL

Operating Characteristic Curves • Single Sampling Plans

Average Outgoing Quality Limit, AOQL = 0.75%

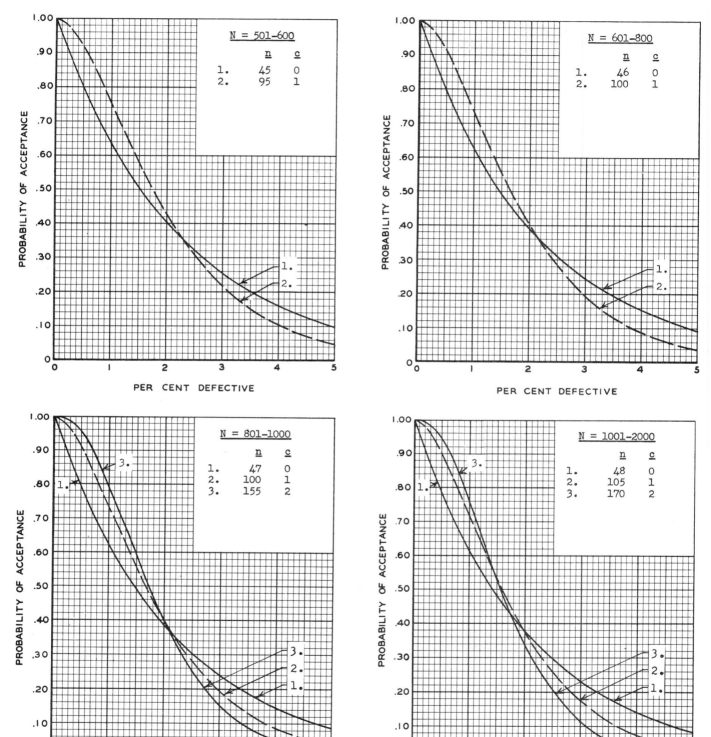

N = 501–600

	n	c
1.	45	0
2.	95	1

N = 601–800

	n	c
1.	46	0
2.	100	1

N = 801–1000

	n	c
1.	47	0
2.	100	1
3.	155	2

N = 1001–2000

	n	c
1.	48	0
2.	105	1
3.	170	2

Operating Characteristic Curves • Single Sampling Plans
Average Outgoing Quality Limit, AOQL = 0.75%

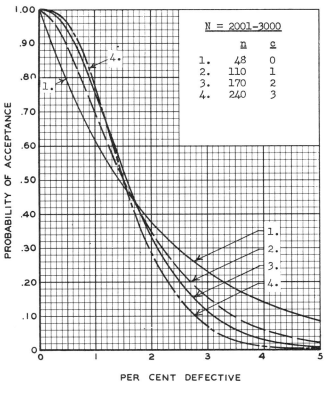

N = 2001–3000

	n	c
1.	48	0
2.	110	1
3.	170	2
4.	240	3

N = 3001–4000

	n	c
1.	48	0
2.	110	1
3.	175	2
4.	245	3
5.	315	4

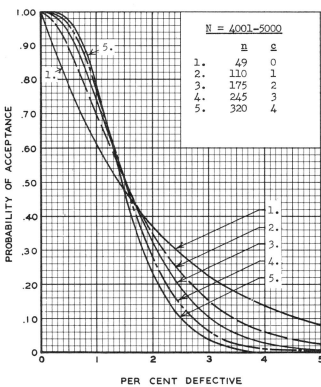

N = 4001–5000

	n	c
1.	49	0
2.	110	1
3.	175	2
4.	245	3
5.	320	4

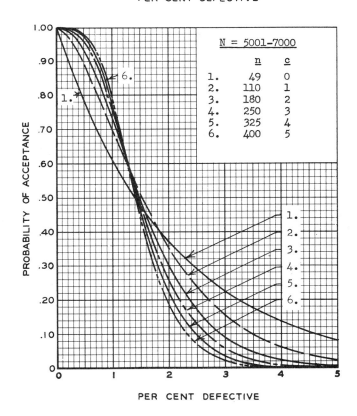

N = 5001–7000

	n	c
1.	49	0
2.	110	1
3.	180	2
4.	250	3
5.	325	4
6.	400	5

SINGLE
SAMPLING

0.75%

AOQL

Operating Characteristic Curves • Single Sampling Plans

Average Outgoing Quality Limit, AOQL = 0.75%

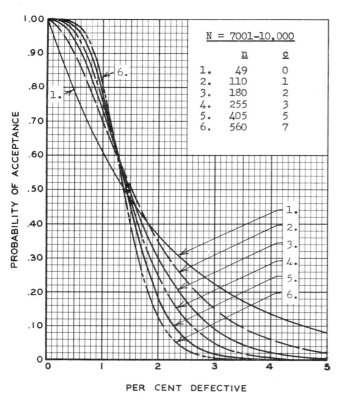

N = 7001-10,000

	n	c
1.	49	0
2.	110	1
3.	180	2
4.	255	3
5.	405	5
6.	560	7

N = 10,001-20,000

	n	c
1.	49	0
2.	110	1
3.	255	3
4.	335	4
5.	495	6
6.	750	9

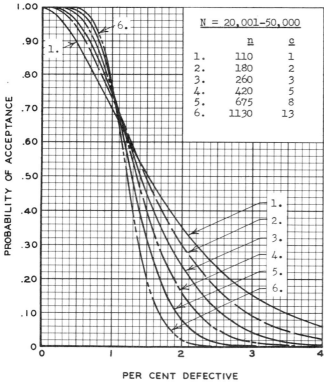

N = 20,001-50,000

	n	c
1.	110	1
2.	180	2
3.	260	3
4.	420	5
5.	675	8
6.	1130	13

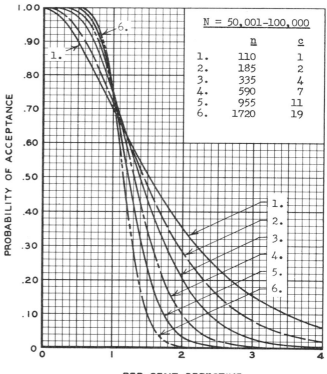

N = 50,001-100,000

	n	c
1.	110	1
2.	185	2
3.	335	4
4.	590	7
5.	955	11
6.	1720	19

Operating Characteristic Curves • Single Sampling Plans
Average Outgoing Quality Limit, AOQL = 1.0%

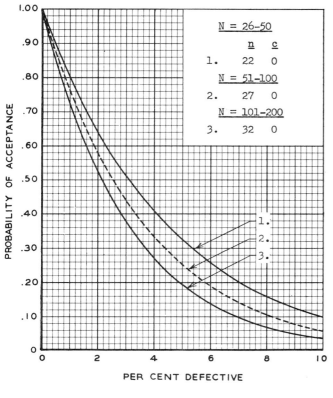

N = 26–50

	n	c
1.	22	0

N = 51–100

	n	c
2.	27	0

N = 101–200

	n	c
3.	32	0

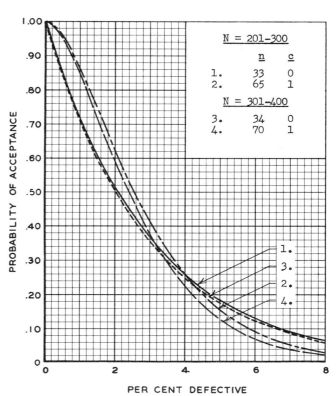

N = 201–300

	n	c
1.	33	0
2.	65	1

N = 301–400

	n	c
3.	34	0
4.	70	1

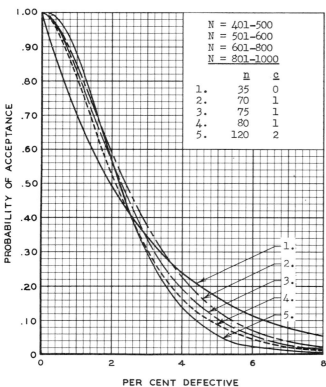

N = 401–500
N = 501–600
N = 601–800
N = 801–1000

	n	c
1.	35	0
2.	70	1
3.	75	1
4.	80	1
5.	120	2

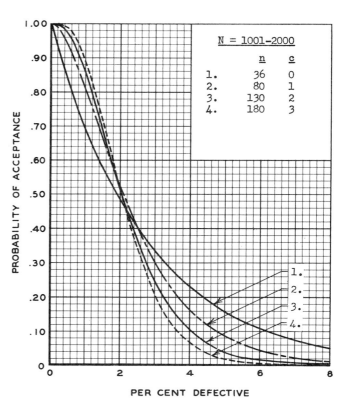

N = 1001–2000

	n	c
1.	36	0
2.	80	1
3.	130	2
4.	180	3

SINGLE
SAMPLING
1.0%
AOQL

Operating Characteristic Curves • Single Sampling Plans
Average Outgoing Quality Limit, AOQL = 1.0%

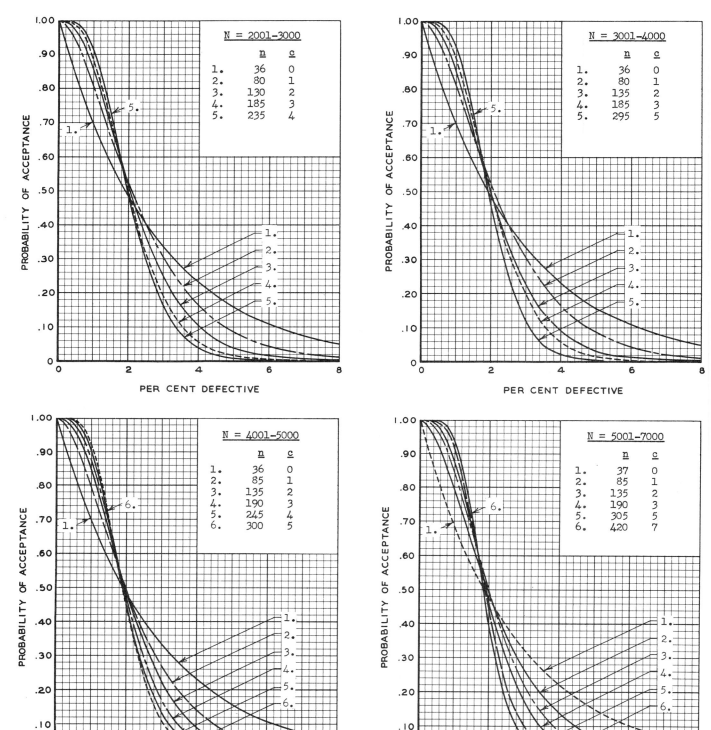

Operating Characteristic Curves • Single Sampling Plans
Average Outgoing Quality Limit, AOQL = 1.0%

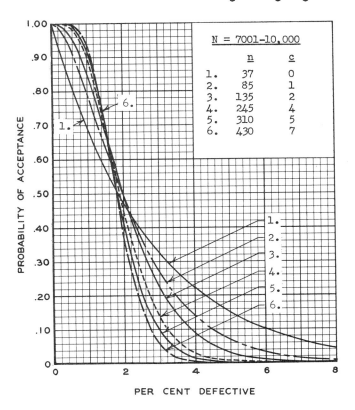

N = 7001-10,000

	n	c
1.	37	0
2.	85	1
3.	135	2
4.	245	4
5.	310	5
6.	430	7

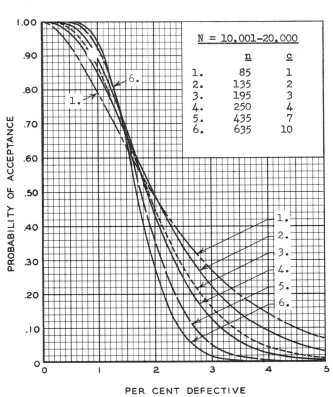

N = 10,001-20,000

	n	c
1.	85	1
2.	135	2
3.	195	3
4.	250	4
5.	435	7
6.	635	10

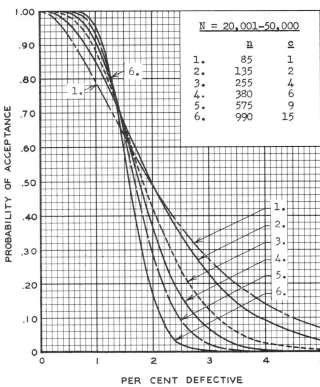

N = 20,001-50,000

	n	c
1.	85	1
2.	135	2
3.	255	4
4.	380	6
5.	575	9
6.	990	15

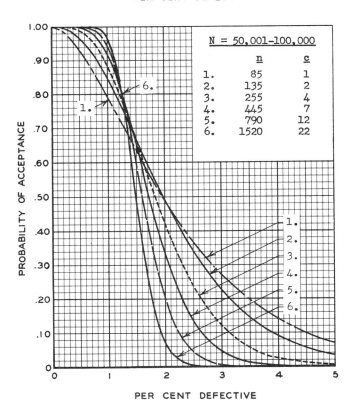

N = 50,001-100,000

	n	c
1.	85	1
2.	135	2
3.	255	4
4.	445	7
5.	790	12
6.	1520	22

SINGLE
SAMPLING

1.5%
AOQL

Operating Characteristic Curves • Single Sampling Plans
Average Outgoing Quality Limit, AOQL = 1.5%

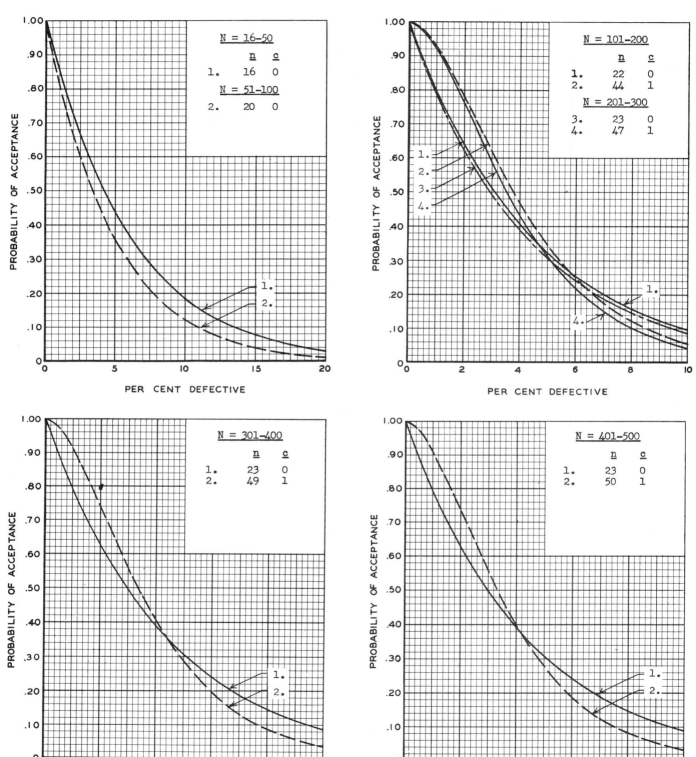

SINGLE
SAMPLING
1.5%
AOQL

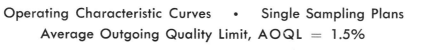

Operating Characteristic Curves • Single Sampling Plans
Average Outgoing Quality Limit, AOQL = 1.5%

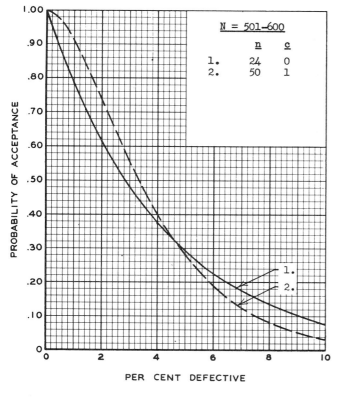

N = 501–600

	n	c
1.	24	0
2.	50	1

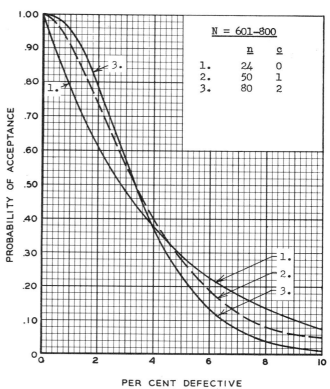

N = 601–800

	n	c
1.	24	0
2.	50	1
3.	80	2

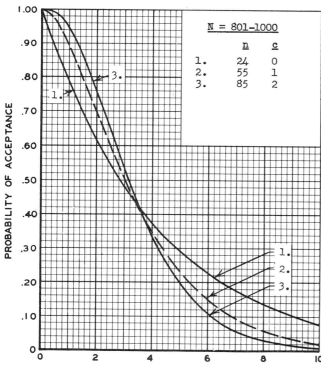

N = 801–1000

	n	c
1.	24	0
2.	55	1
3.	85	2

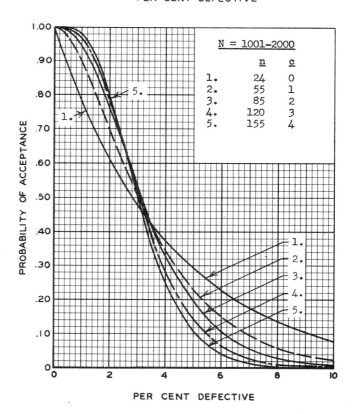

N = 1001–2000

	n	c
1.	24	0
2.	55	1
3.	85	2
4.	120	3
5.	155	4

SINGLE SAMPLING

1.5%

AOQL

Operating Characteristic Curves • Single Sampling Plans
Average Outgoing Quality Limit, AOQL = 1.5%

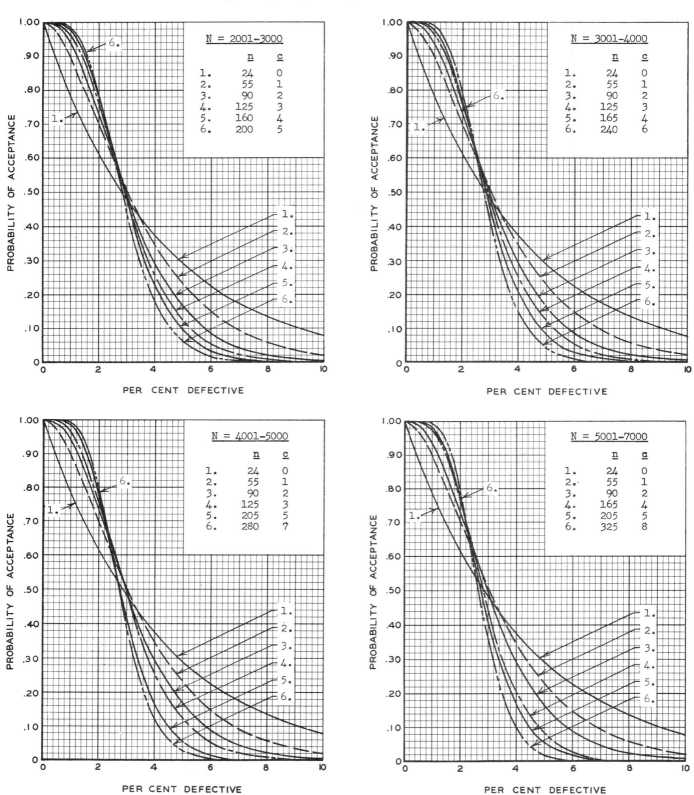

N = 2001–3000

	n	c
1.	24	0
2.	55	1
3.	90	2
4.	125	3
5.	160	4
6.	200	5

N = 3001–4000

	n	c
1.	24	0
2.	55	1
3.	90	2
4.	125	3
5.	165	4
6.	240	6

N = 4001–5000

	n	c
1.	24	0
2.	55	1
3.	90	2
4.	125	3
5.	205	5
6.	280	7

N = 5001–7000

	n	c
1.	24	0
2.	55	1
3.	90	2
4.	165	4
5.	205	5
6.	325	8

PER CENT DEFECTIVE

PROBABILITY OF ACCEPTANCE

Operating Characteristic Curves • Single Sampling Plans
Average Outgoing Quality Limit, AOQL = 1.5%

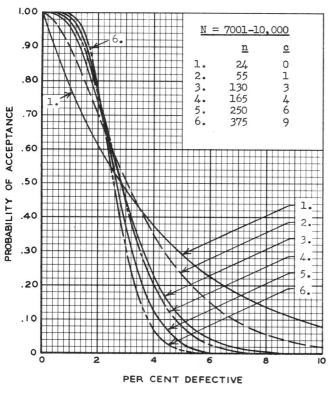

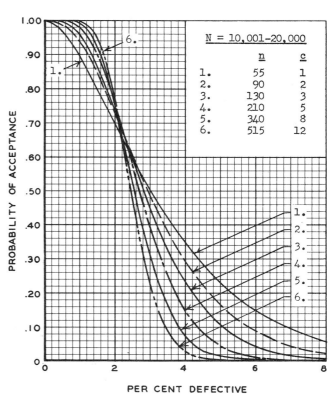

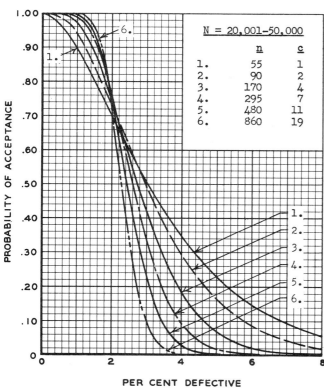

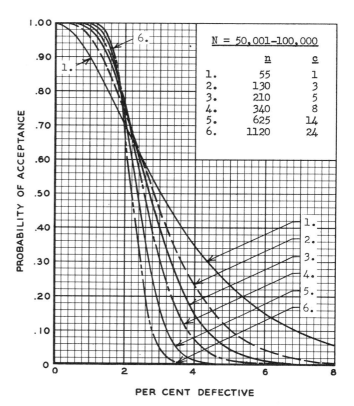

SINGLE
SAMPLING
2.0%
AOQL

Operating Characteristic Curves • Single Sampling Plans
Average Outgoing Quality Limit, AOQL = 2.0%

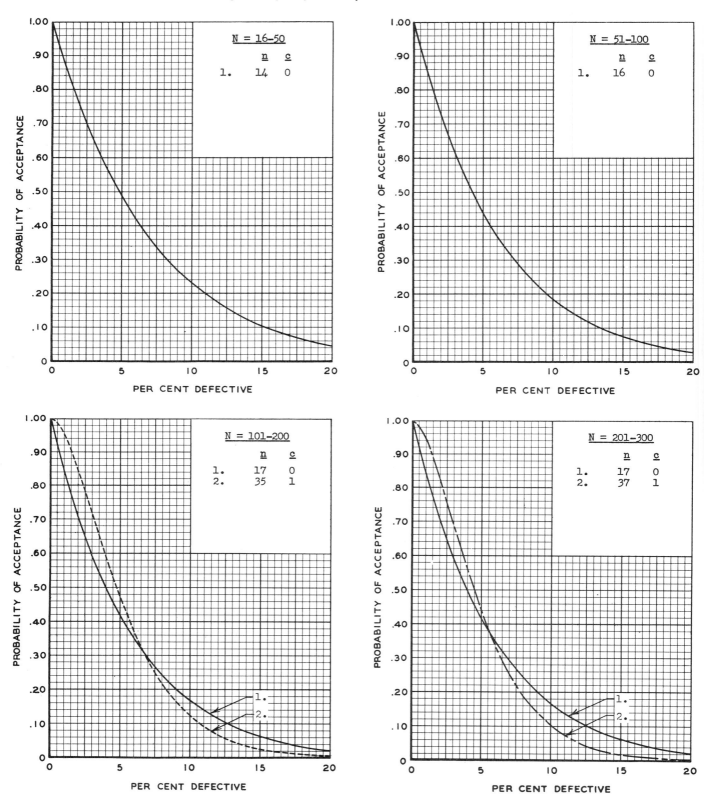

Operating Characteristic Curves • Single Sampling Plans
Average Outgoing Quality Limit, AOQL = 2.0%

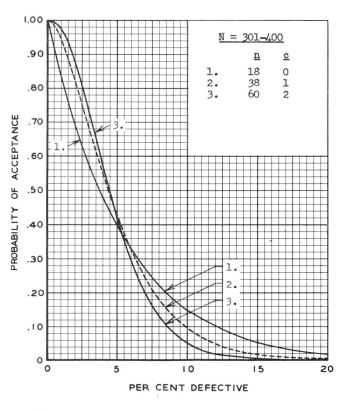

N = 301–400

	n	c
1.	18	0
2.	38	1
3.	60	2

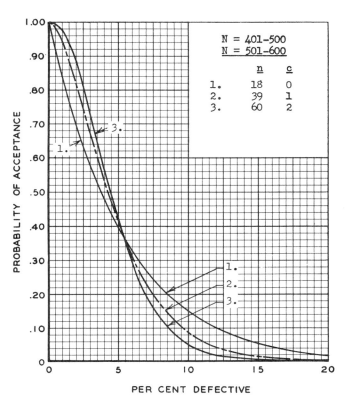

N = 401–500
N = 501–600

	n	c
1.	18	0
2.	39	1
3.	60	2

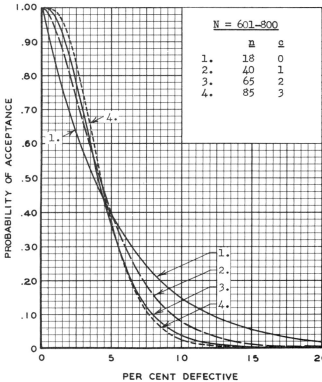

N = 601–800

	n	c
1.	18	0
2.	40	1
3.	65	2
4.	85	3

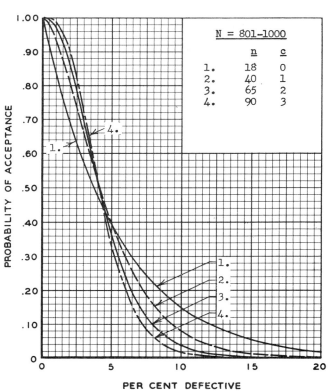

N = 801–1000

	n	c
1.	18	0
2.	40	1
3.	65	2
4.	90	3

SINGLE
SAMPLING
2.0%
AOQL

Operating Characteristic Curves　•　Single Sampling Plans
Average Outgoing Quality Limit, AOQL = 2.0%

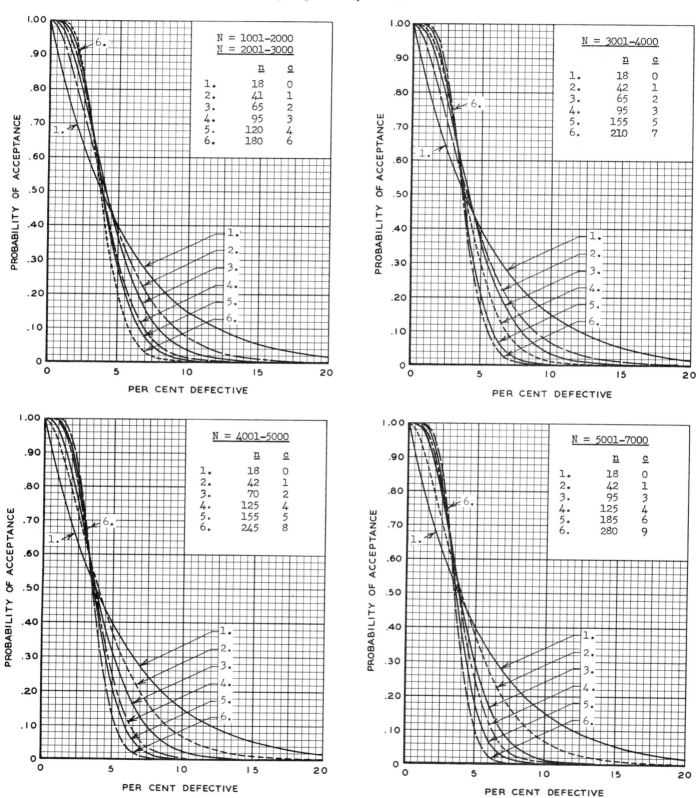

Top-left chart: N = 1001–2000, N = 2001–3000

	n	c
1.	18	0
2.	41	1
3.	65	2
4.	95	3
5.	120	4
6.	180	6

Top-right chart: N = 3001–4000

	n	c
1.	18	0
2.	42	1
3.	65	2
4.	95	3
5.	155	5
6.	210	7

Bottom-left chart: N = 4001–5000

	n	c
1.	18	0
2.	42	1
3.	70	2
4.	125	4
5.	155	5
6.	245	8

Bottom-right chart: N = 5001–7000

	n	c
1.	18	0
2.	42	1
3.	95	3
4.	125	4
5.	185	6
6.	280	9

All charts: axes labeled PROBABILITY OF ACCEPTANCE (vertical) and PER CENT DEFECTIVE (horizontal).

Operating Characteristic Curves • Single Sampling Plans

Average Outgoing Quality Limit, AOQL = 2.0%

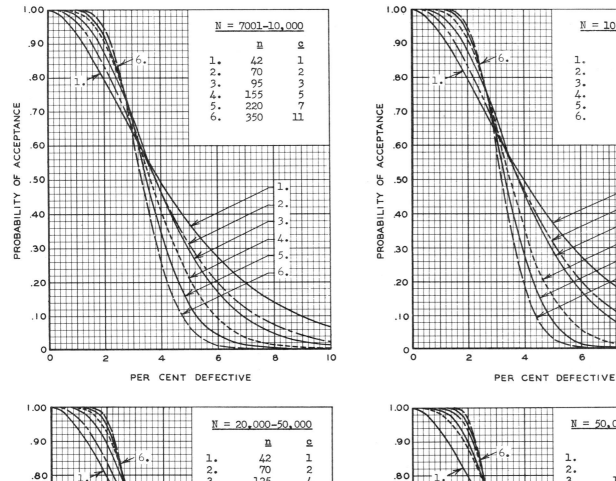

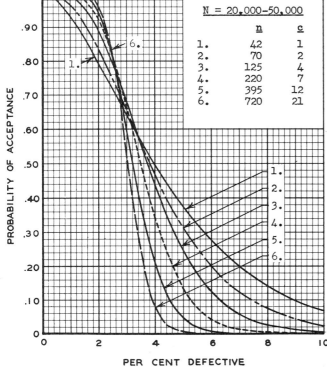

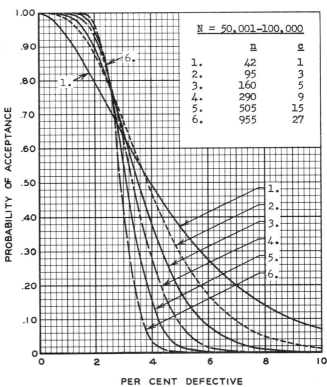

SINGLE
SAMPLING
2.5%
AOQL

Operating Characteristic Curves • Single Sampling Plans
Average Outgoing Quality Limit, AOQL = 2.5%

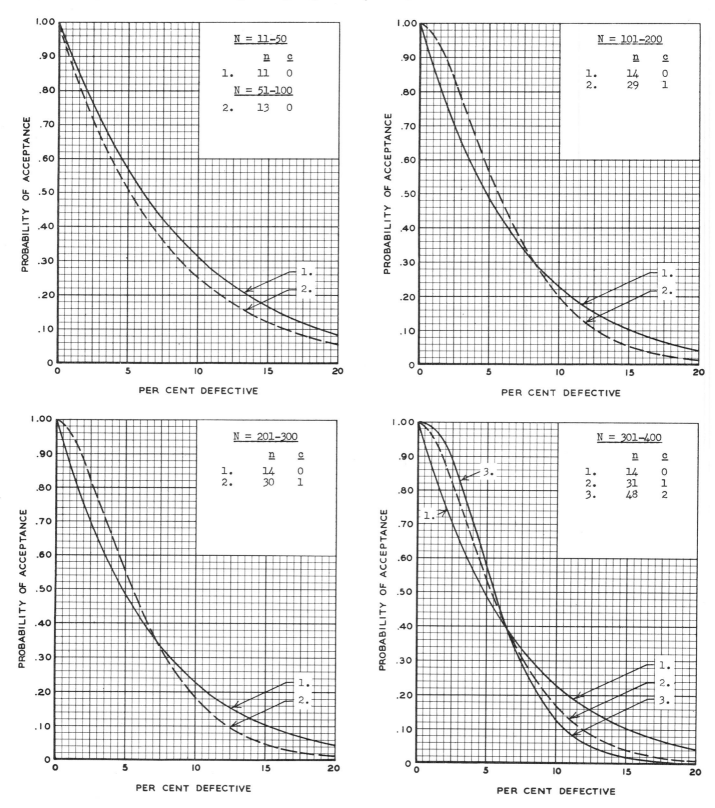

Operating Characteristic Curves • Single Sampling Plans
Average Outgoing Quality Limit, AOQL = 2.5%

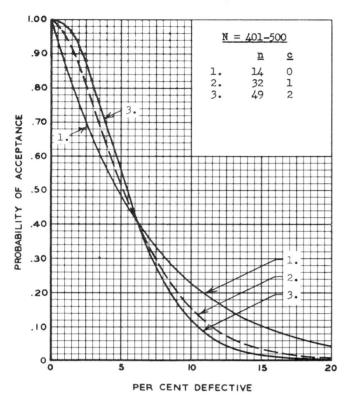

N = 401–500

	n	c
1.	14	0
2.	32	1
3.	49	2

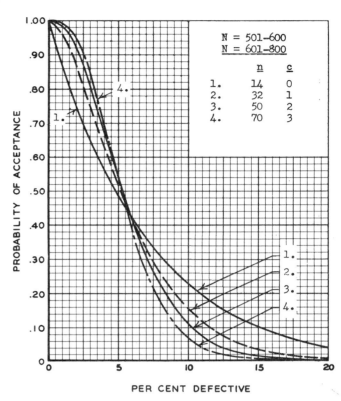

N = 501–600
N = 601–800

	n	c
1.	14	0
2.	32	1
3.	50	2
4.	70	3

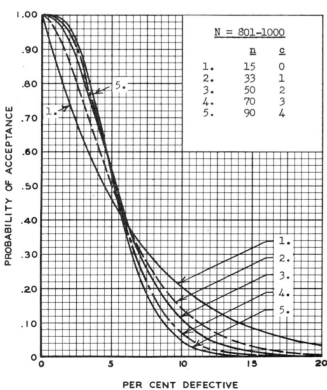

N = 801–1000

	n	c
1.	15	0
2.	33	1
3.	50	2
4.	70	3
5.	90	4

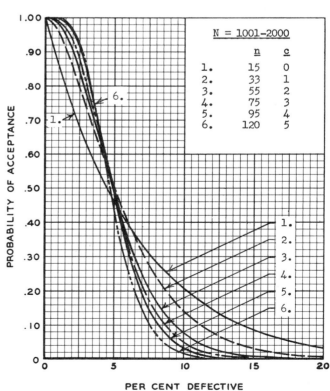

N = 1001–2000

	n	c
1.	15	0
2.	33	1
3.	55	2
4.	75	3
5.	95	4
6.	120	5

SINGLE
SAMPLING
2.5%
AOQL

Operating Characteristic Curves • Single Sampling Plans
Average Outgoing Quality Limit, AOQL = 2.5%

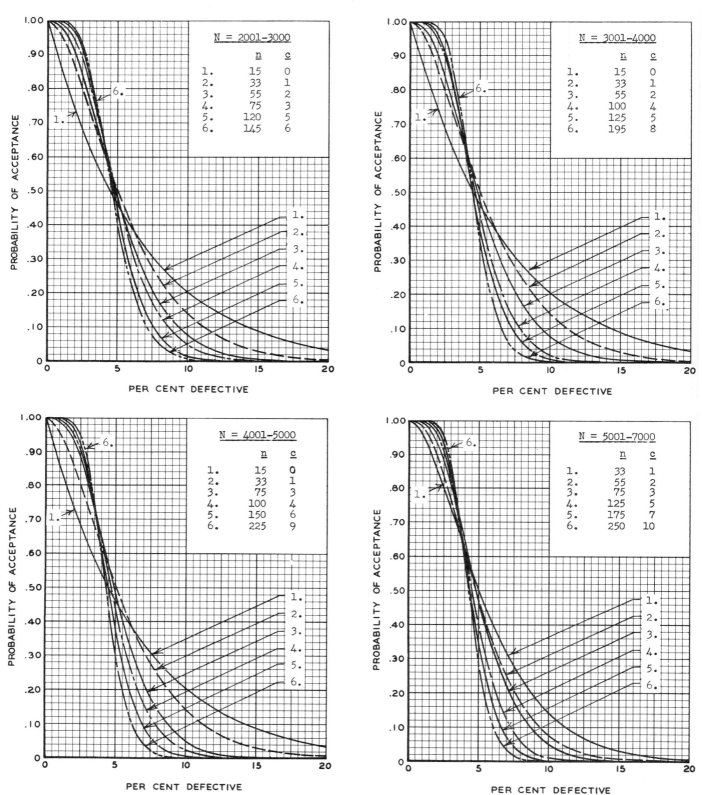

N = 2001–3000

	n	c
1.	15	0
2.	33	1
3.	55	2
4.	75	3
5.	120	5
6.	145	6

N = 3001–4000

	n	c
1.	15	0
2.	33	1
3.	55	2
4.	100	4
5.	125	5
6.	195	8

N = 4001–5000

	n	c
1.	15	0
2.	33	1
3.	75	3
4.	100	4
5.	150	6
6.	225	9

N = 5001–7000

	n	c
1.	33	1
2.	55	2
3.	75	3
4.	125	5
5.	175	7
6.	250	10

PROBABILITY OF ACCEPTANCE

PER CENT DEFECTIVE

Operating Characteristic Curves • Single Sampling Plans

Average Outgoing Quality Limit, AOQL = 2.5%

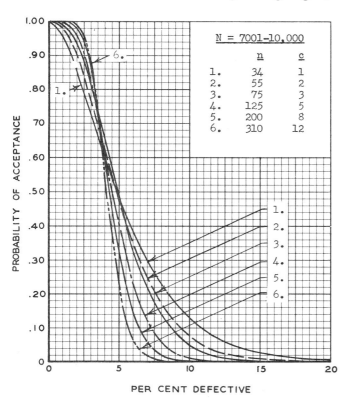

N = 7001–10,000

	n	c
1.	34	1
2.	55	2
3.	75	3
4.	125	5
5.	200	8
6.	310	12

N = 10,001–20,000

	n	c
1.	34	1
2.	55	2
3.	100	4
4.	150	6
5.	260	10
6.	425	16

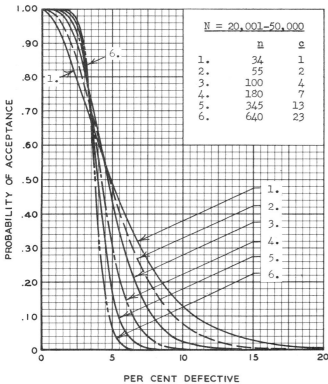

N = 20,001–50,000

	n	c
1.	34	1
2.	55	2
3.	100	4
4.	180	7
5.	345	13
6.	640	23

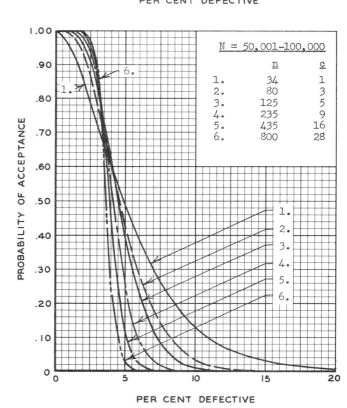

N = 50,001–100,000

	n	c
1.	34	1
2.	80	3
3.	125	5
4.	235	9
5.	435	16
6.	800	28

SINGLE
SAMPLING
3.0%
AOQL

Operating Characteristic Curves • Single Sampling Plans
Average Outgoing Quality Limit, AOQL = 3.0%

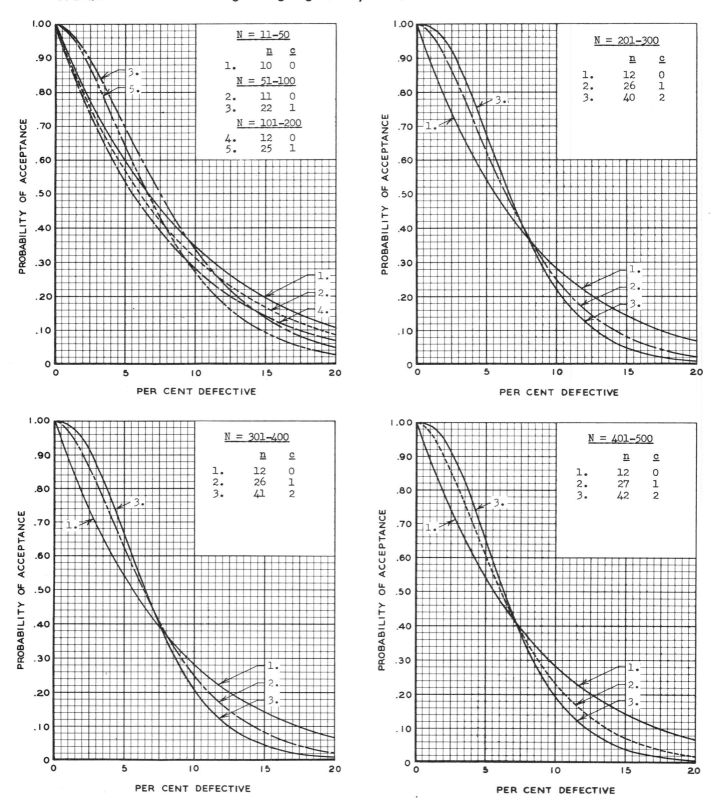

Operating Characteristic Curves • Single Sampling Plans
Average Outgoing Quality Limit, AOQL = 3.0%

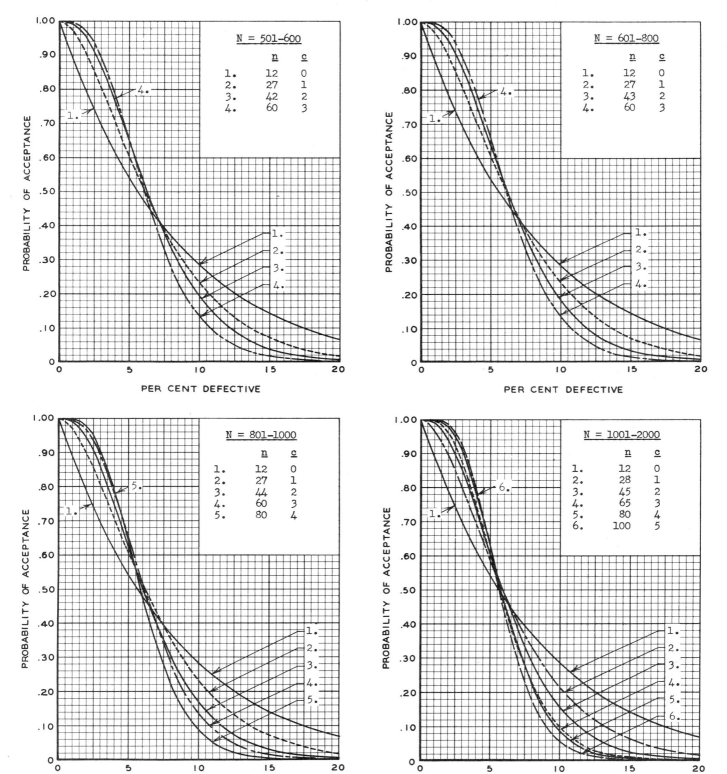

SINGLE
SAMPLING

3.0%

AOQL

Operating Characteristic Curves • Single Sampling Plans
Average Outgoing Quality Limit, AOQL = 3.0%

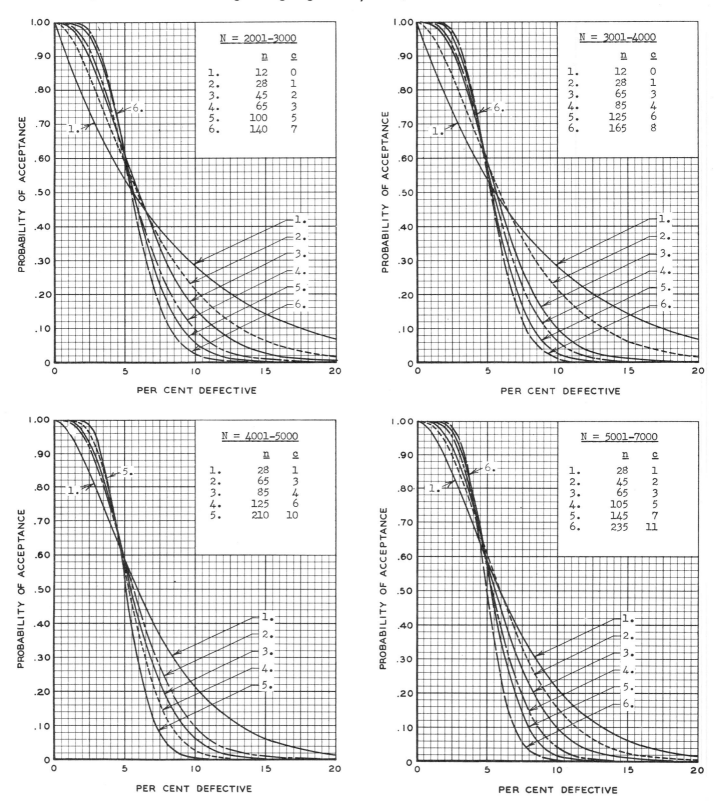

Operating Characteristic Curves • Single Sampling Plans
Average Outgoing Quality Limit, AOQL = 3.0%

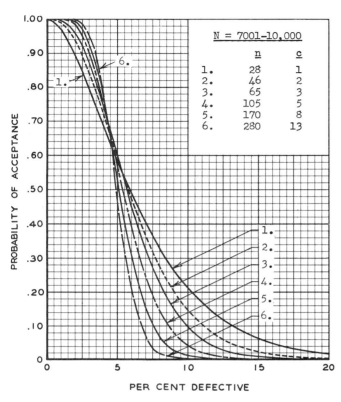

N = 7001–10,000

	n	c
1.	28	1
2.	46	2
3.	65	3
4.	105	5
5.	170	8
6.	280	13

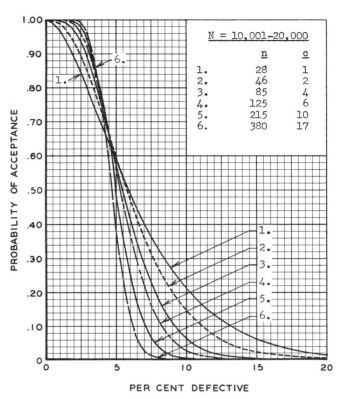

N = 10,001–20,000

	n	c
1.	28	1
2.	46	2
3.	85	4
4.	125	6
5.	215	10
6.	380	17

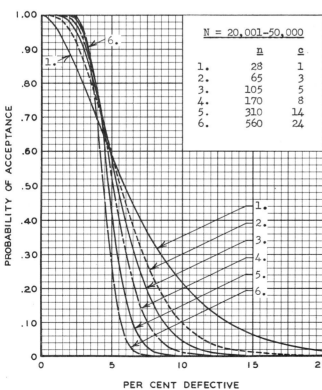

N = 20,001–50,000

	n	c
1.	28	1
2.	65	3
3.	105	5
4.	170	8
5.	310	14
6.	560	24

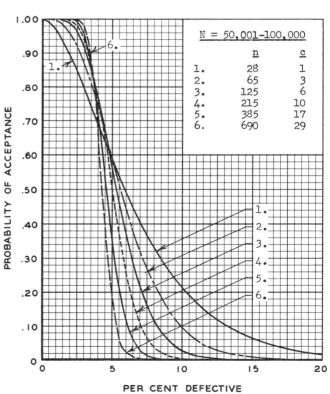

N = 50,001–100,000

	n	c
1.	28	1
2.	65	3
3.	125	6
4.	215	10
5.	385	17
6.	690	29

SINGLE
SAMPLING
4.0%
AOQL

Operating Characteristic Curves • Single Sampling Plans
Average Outgoing Quality Limit, AOQL = 4.0%

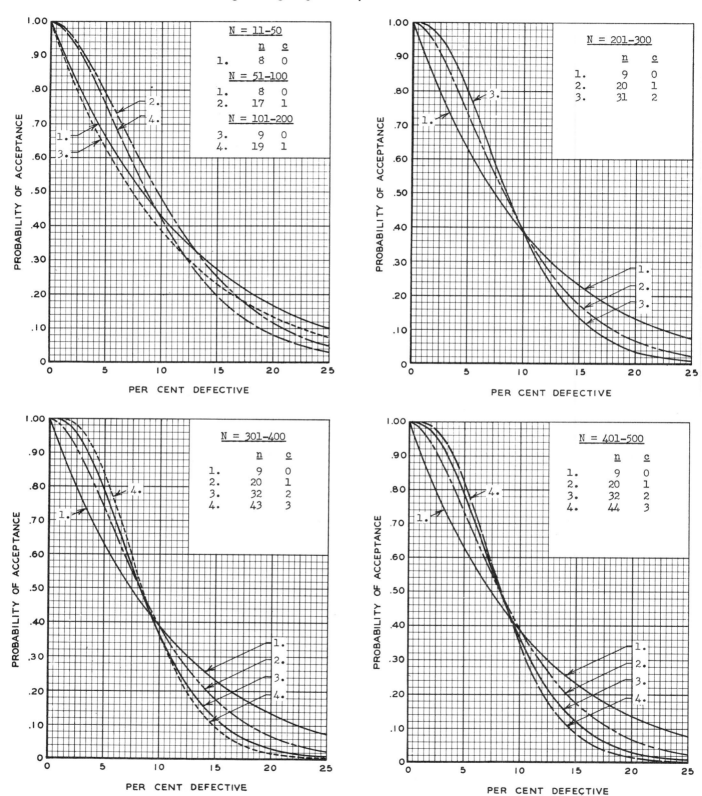

Operating Characteristic Curves • Single Sampling Plans
Average Outgoing Quality Limit, AOQL = 4.0%

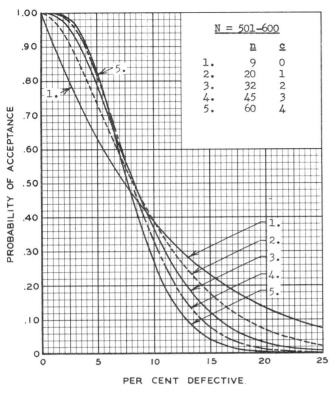

N = 501–600

	n	c
1.	9	0
2.	20	1
3.	32	2
4.	45	3
5.	60	4

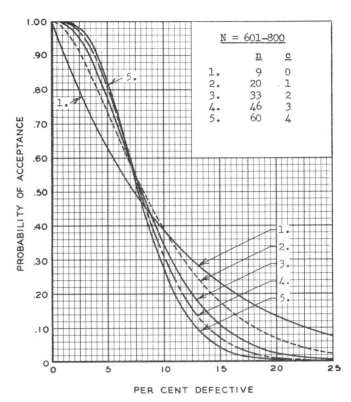

N = 601–800

	n	c
1.	9	0
2.	20	1
3.	33	2
4.	46	3
5.	60	4

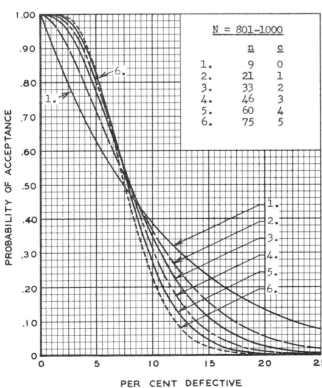

N = 801–1000

	n	c
1.	9	0
2.	21	1
3.	33	2
4.	46	3
5.	60	4
6.	75	5

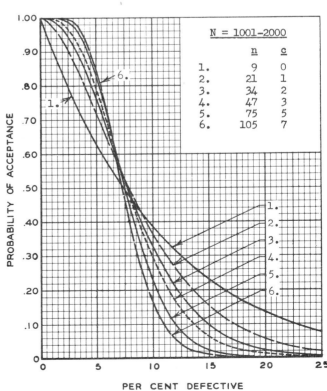

N = 1001–2000

	n	c
1.	9	0
2.	21	1
3.	34	2
4.	47	3
5.	75	5
6.	105	7

SINGLE
SAMPLING
4.0%
AOQL

Operating Characteristic Curves • Single Sampling Plans
Average Outgoing Quality Limit, AOQL = 4.0%

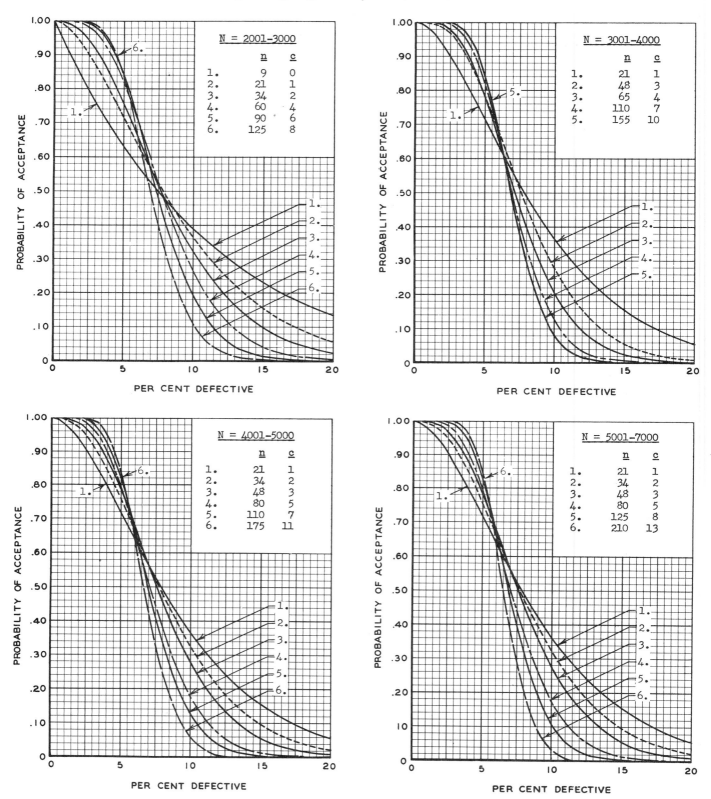

Operating Characteristic Curves • Single Sampling Plans
Average Outgoing Quality Limit, AOQL = 4.0%

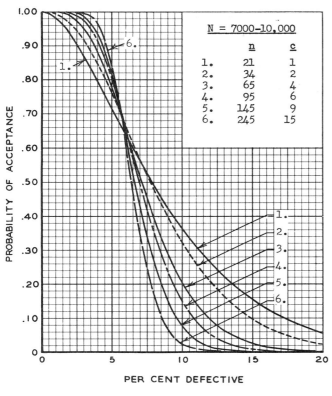

N = 7000-10,000

	n	c
1.	21	1
2.	34	2
3.	65	4
4.	95	6
5.	145	9
6.	245	15

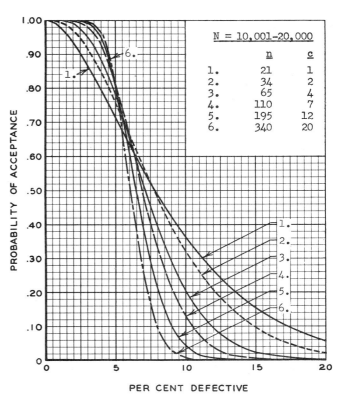

N = 10,001-20,000

	n	c
1.	21	1
2.	34	2
3.	65	4
4.	110	7
5.	195	12
6.	340	20

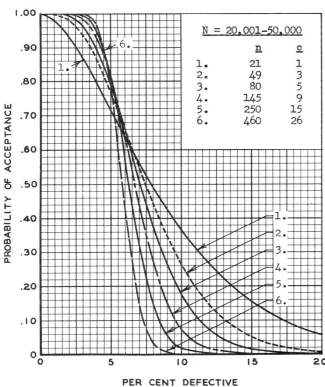

N = 20,001-50,000

	n	c
1.	21	1
2.	49	3
3.	80	5
4.	145	9
5.	250	15
6.	460	26

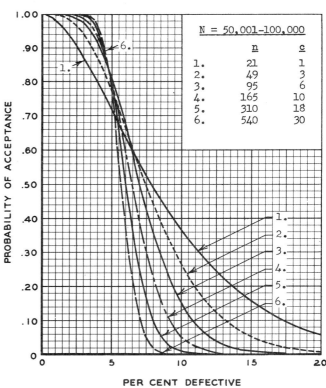

N = 50,001-100,000

	n	c
1.	21	1
2.	49	3
3.	95	6
4.	165	10
5.	310	18
6.	540	30

SINGLE
SAMPLING
5.0%
AOQL

Operating Characteristic Curves • Single Sampling Plans
Average Outgoing Quality Limit, AOQL = 5.0%

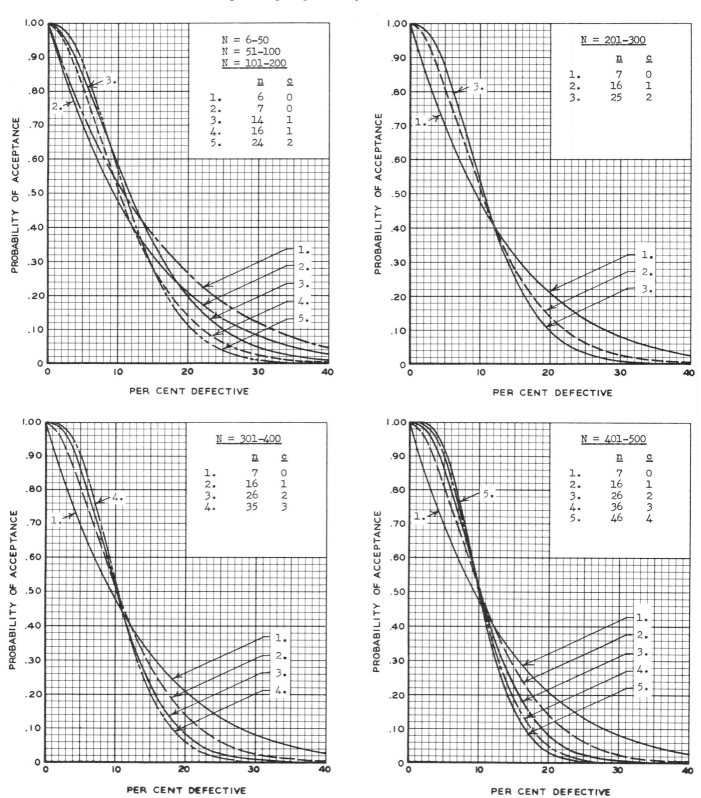

Operating Characteristic Curves • Single Sampling Plans
Average Outgoing Quality Limit, AOQL = 5.0%

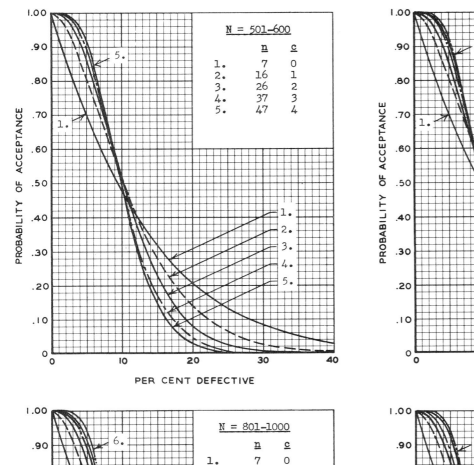

N = 501-600

	n	c
1.	7	0
2.	16	1
3.	26	2
4.	37	3
5.	47	4

PROBABILITY OF ACCEPTANCE

PER CENT DEFECTIVE

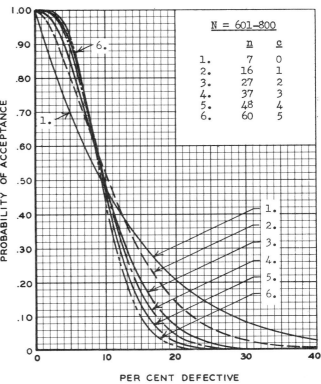

N = 601-800

	n	c
1.	7	0
2.	16	1
3.	27	2
4.	37	3
5.	48	4
6.	60	5

PROBABILITY OF ACCEPTANCE

PER CENT DEFECTIVE

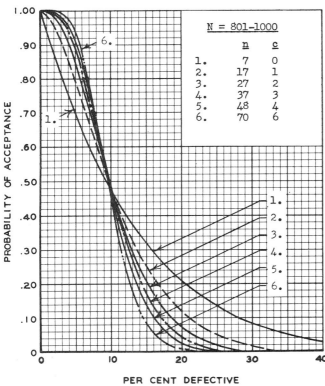

N = 801-1000

	n	c
1.	7	0
2.	17	1
3.	27	2
4.	37	3
5.	48	4
6.	70	6

PROBABILITY OF ACCEPTANCE

PER CENT DEFECTIVE

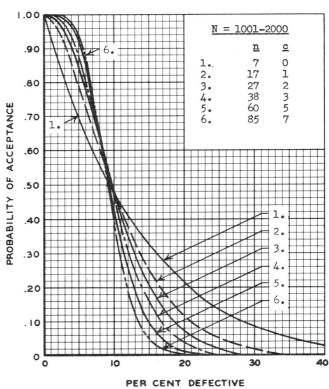

N = 1001-2000

	n	c
1.	7	0
2.	17	1
3.	27	2
4.	38	3
5.	60	5
6.	85	7

PROBABILITY OF ACCEPTANCE

PER CENT DEFECTIVE

SINGLE
SAMPLING
5.0%
AOQL

Operating Characteristic Curves • Single Sampling Plans

Average Outgoing Quality Limit, AOQL = 5.0%

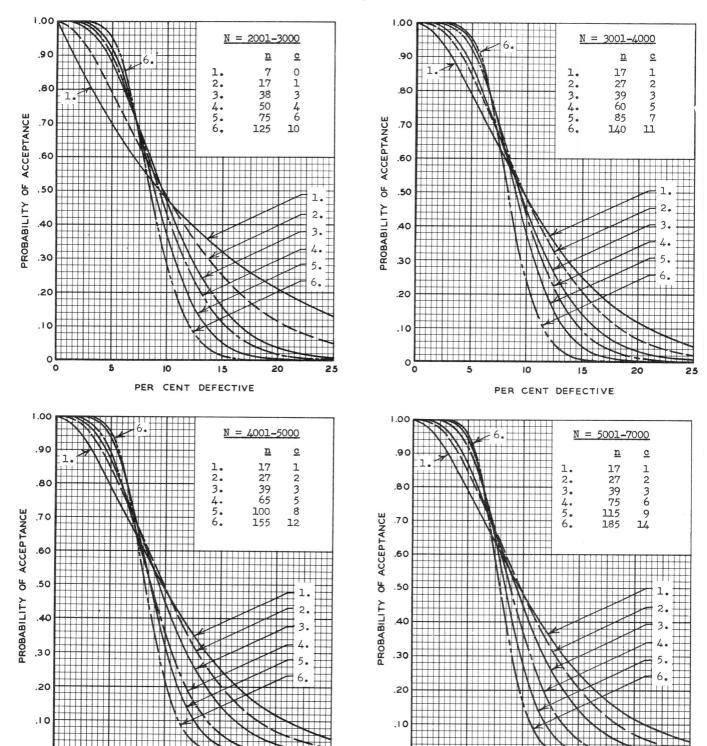

PROBABILITY OF ACCEPTANCE — PER CENT DEFECTIVE

N = 2001–3000

	n	c
1.	7	0
2.	17	1
3.	38	3
4.	50	4
5.	75	6
6.	125	10

N = 3001–4000

	n	c
1.	17	1
2.	27	2
3.	39	3
4.	60	5
5.	85	7
6.	140	11

N = 4001–5000

	n	c
1.	17	1
2.	27	2
3.	39	3
4.	65	5
5.	100	8
6.	155	12

N = 5001–7000

	n	c
1.	17	1
2.	27	2
3.	39	3
4.	75	6
5.	115	9
6.	185	14

Operating Characteristic Curves • Single Sampling Plans
Average Outgoing Quality Limit, AOQL = 5.0%

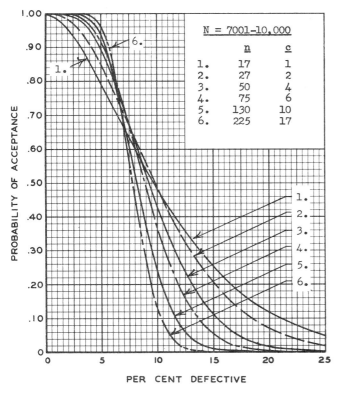

N = 7001-10,000

	n	c
1.	17	1
2.	27	2
3.	50	4
4.	75	6
5.	130	10
6.	225	17

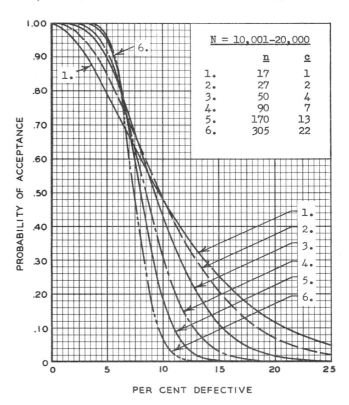

N = 10,001-20,000

	n	c
1.	17	1
2.	27	2
3.	50	4
4.	90	7
5.	170	13
6.	305	22

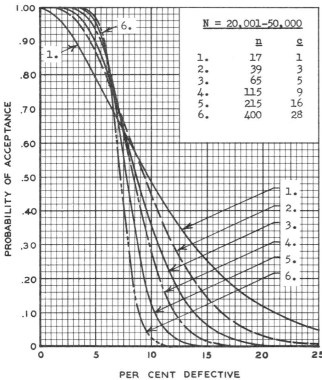

N = 20,001-50,000

	n	c
1.	17	1
2.	39	3
3.	65	5
4.	115	9
5.	215	16
6.	400	28

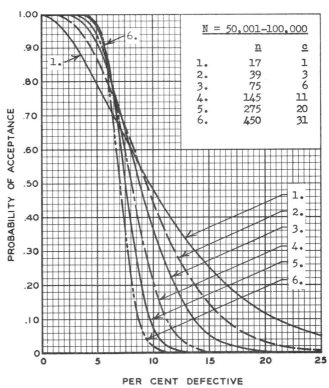

N = 50,001-100,000

	n	c
1.	17	1
2.	39	3
3.	75	6
4.	145	11
5.	275	20
6.	450	31

SINGLE
SAMPLING
7.0%
AOQL

Operating Characteristic Curves • Single Sampling Plans
Average Outgoing Quality Limit, AOQL = 7.0%

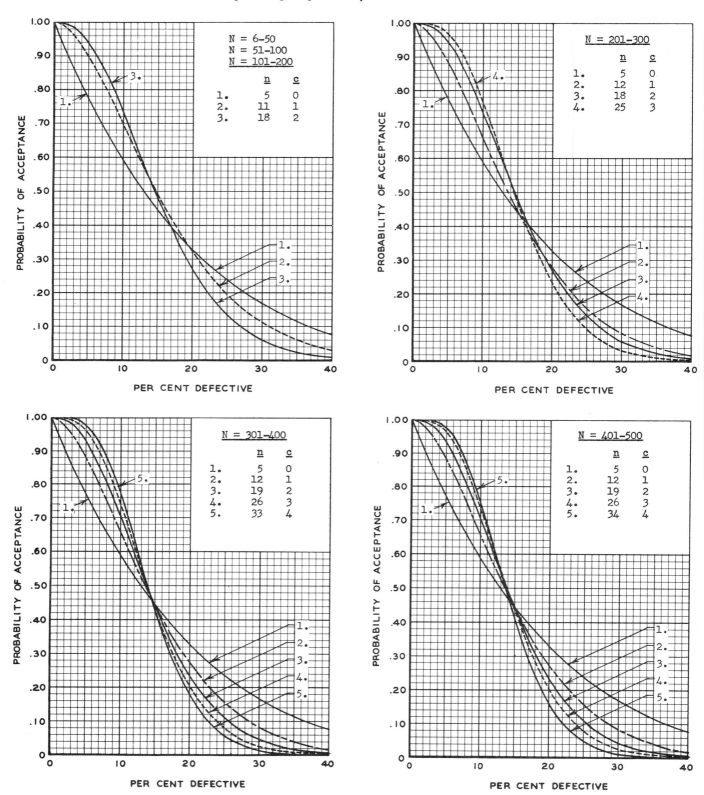

Operating Characteristic Curves • Single Sampling Plans

Average Outgoing Quality Limit, AOQL = 7.0%

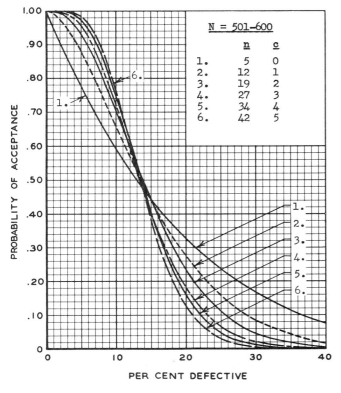

N = 501–600

	n	c
1.	5	0
2.	12	1
3.	19	2
4.	27	3
5.	34	4
6.	42	5

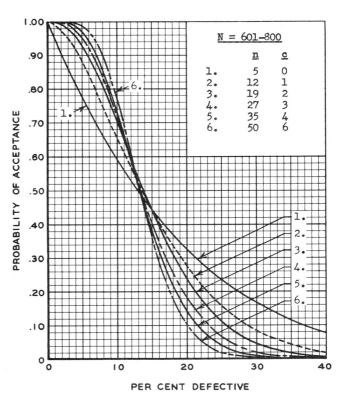

N = 601–800

	n	c
1.	5	0
2.	12	1
3.	19	2
4.	27	3
5.	35	4
6.	50	6

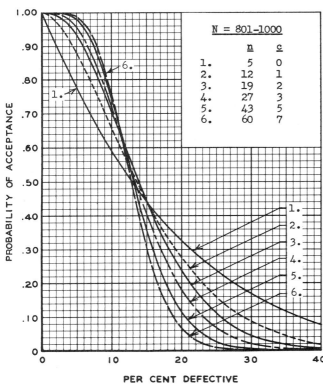

N = 801–1000

	n	c
1.	5	0
2.	12	1
3.	19	2
4.	27	3
5.	43	5
6.	60	7

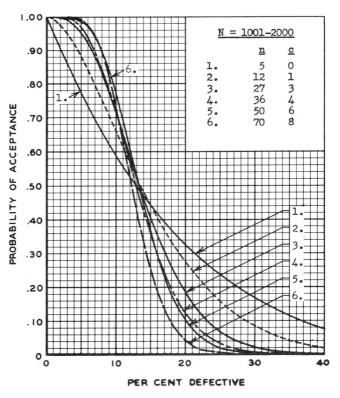

N = 1001–2000

	n	c
1.	5	0
2.	12	1
3.	27	3
4.	36	4
5.	50	6
6.	70	8

SINGLE
SAMPLING

7.0%

AOQL

— 104 —

Appendix 1

Operating Characteristic Curves • Single Sampling Plans
Average Outgoing Quality Limit, AOQL = 7.0%

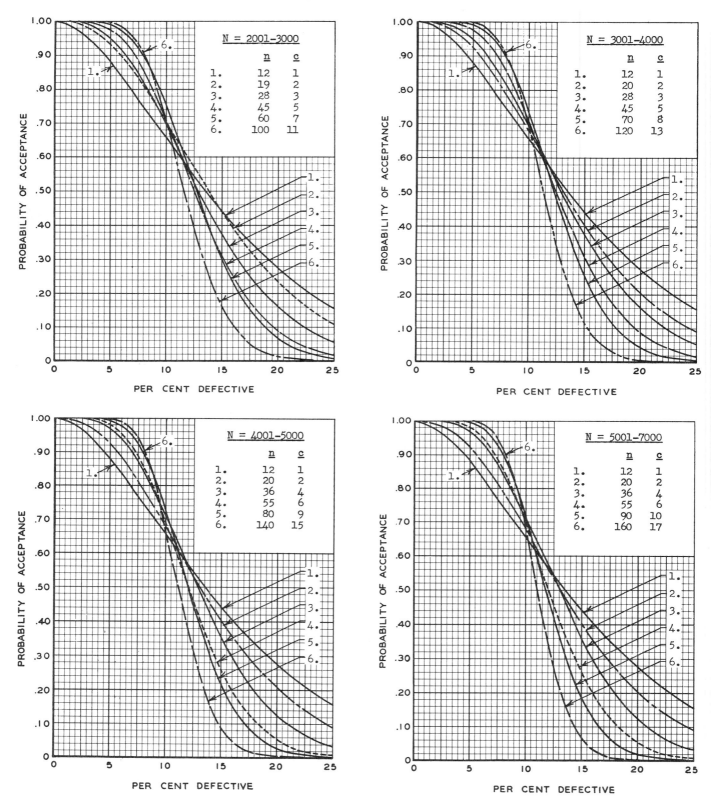

Operating Characteristic Curves • Single Sampling Plans
Average Outgoing Quality Limit, AOQL = 7.0%

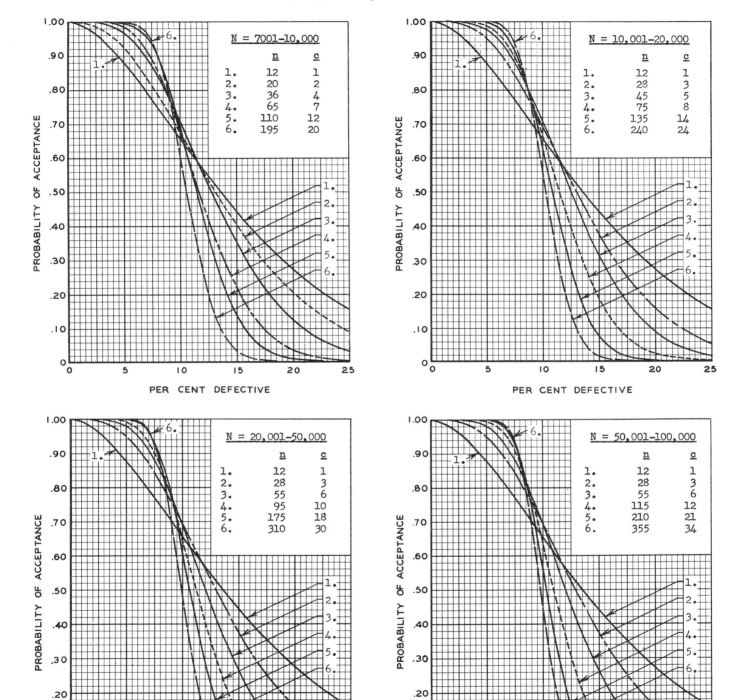

N = 7001-10,000

	n	c
1.	12	1
2.	20	2
3.	36	4
4.	65	7
5.	110	12
6.	195	20

N = 10,001-20,000

	n	c
1.	12	1
2.	28	3
3.	45	5
4.	75	8
5.	135	14
6.	240	24

N = 20,001-50,000

	n	c
1.	12	1
2.	28	3
3.	55	6
4.	95	10
5.	175	18
6.	310	30

N = 50,001-100,000

	n	c
1.	12	1
2.	28	3
3.	55	6
4.	115	12
5.	210	21
6.	355	34

PROBABILITY OF ACCEPTANCE

PER CENT DEFECTIVE

SINGLE
SAMPLING

10.0%
AOQL

Operating Characteristic Curves • Single Sampling Plans
Average Outgoing Quality Limit, AOQL = 10.0%

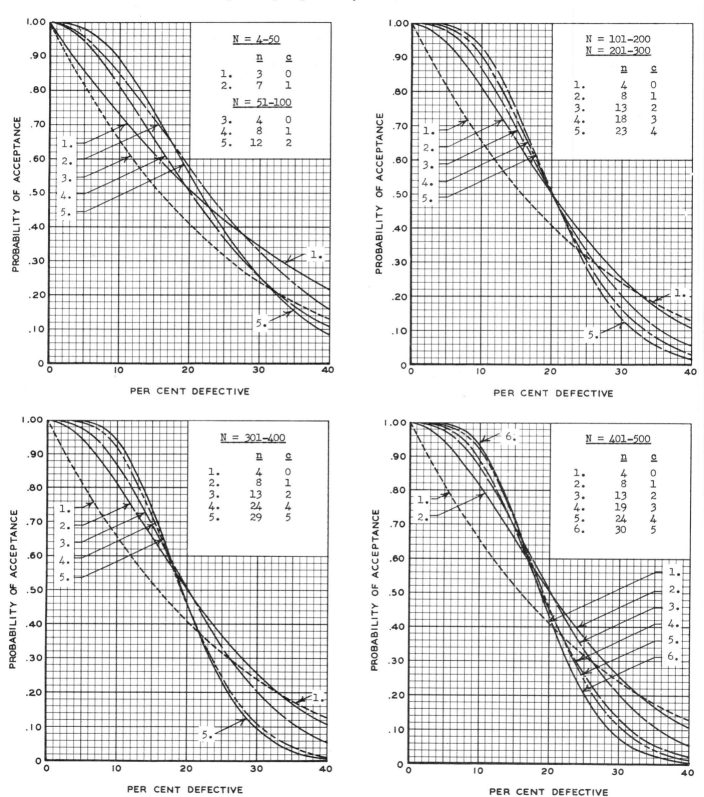

Operating Characteristic Curves • Single Sampling Plans
Average Outgoing Quality Limit, AOQL = 10.0%

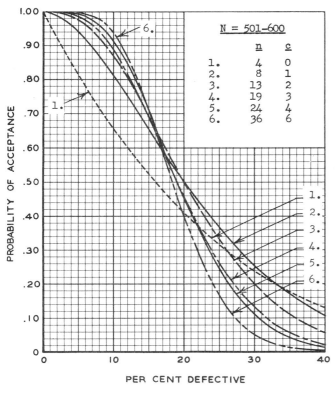

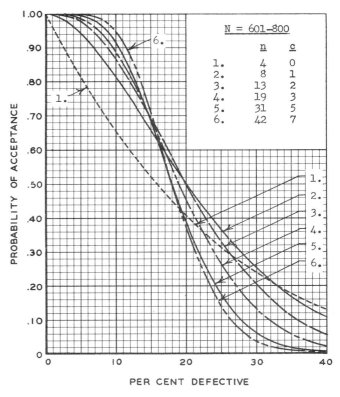

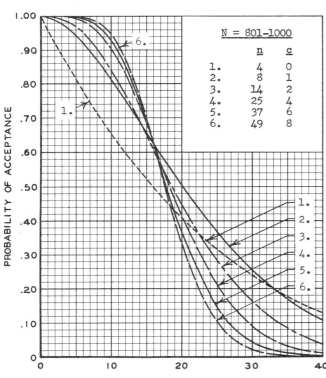

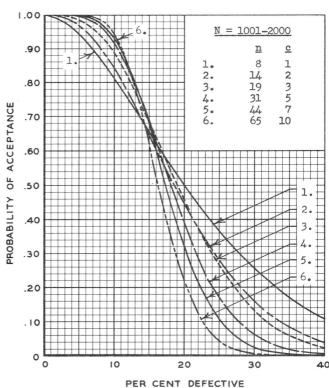

SINGLE SAMPLING

10.0%
AOQL

Operating Characteristic Curves • Single Sampling Plans
Average Outgoing Quality Limit, AOQL = 10.0%

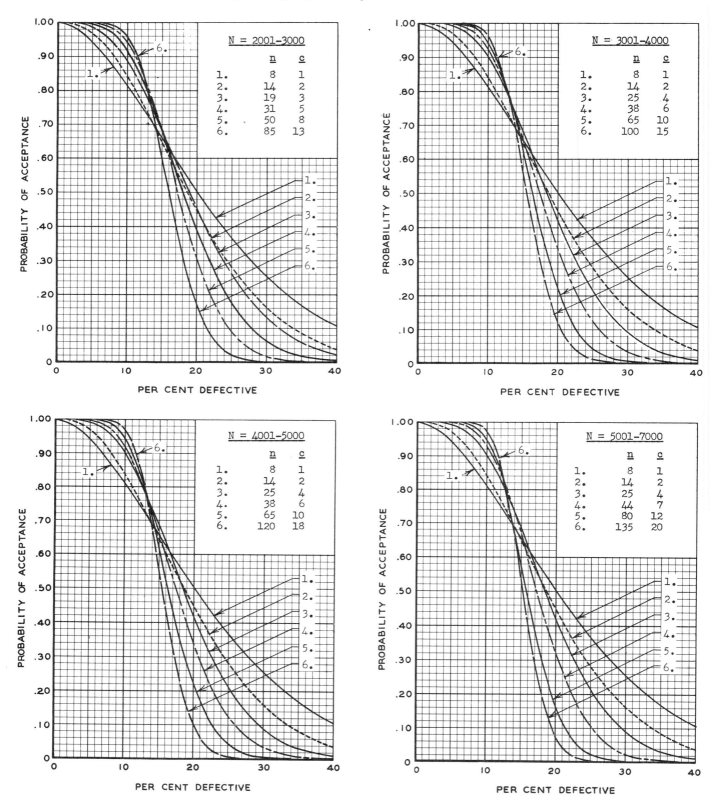

Operating Characteristic Curves • Single Sampling Plans
Average Outgoing Quality Limit, AOQL = 10.0%

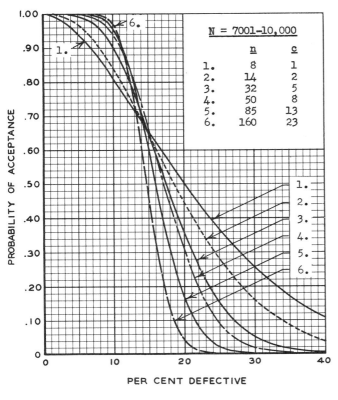

N = 7001-10,000

	n	c
1.	8	1
2.	14	2
3.	32	5
4.	50	8
5.	85	13
6.	160	23

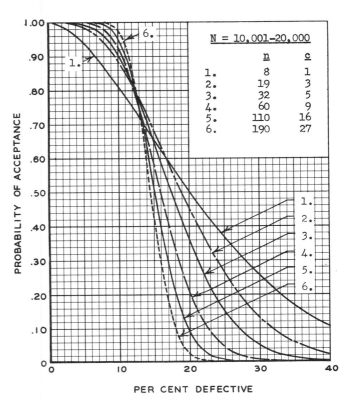

N = 10,001-20,000

	n	c
1.	8	1
2.	19	3
3.	32	5
4.	60	9
5.	110	16
6.	190	27

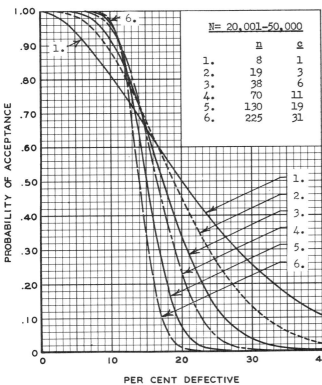

N= 20,001-50,000

	n	c
1.	8	1
2.	19	3
3.	38	6
4.	70	11
5.	130	19
6.	225	31

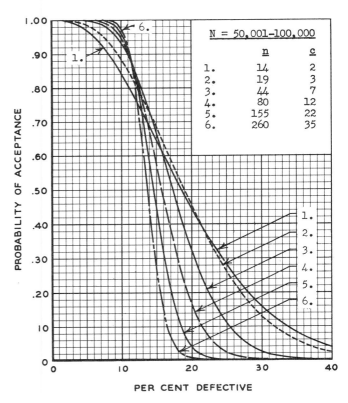

N = 50,001-100,000

	n	c
1.	14	2
2.	19	3
3.	44	7
4.	80	12
5.	155	22
6.	260	35

Appendix 2

OC Curves for All

Double Sampling Plans

in Appendix 7

DOUBLE
SAMPLING

0.1%
AOQL

Operating Characteristic Curves • Double Sampling Plans
Average Outgoing Quality Limit, AOQL = 0.1%

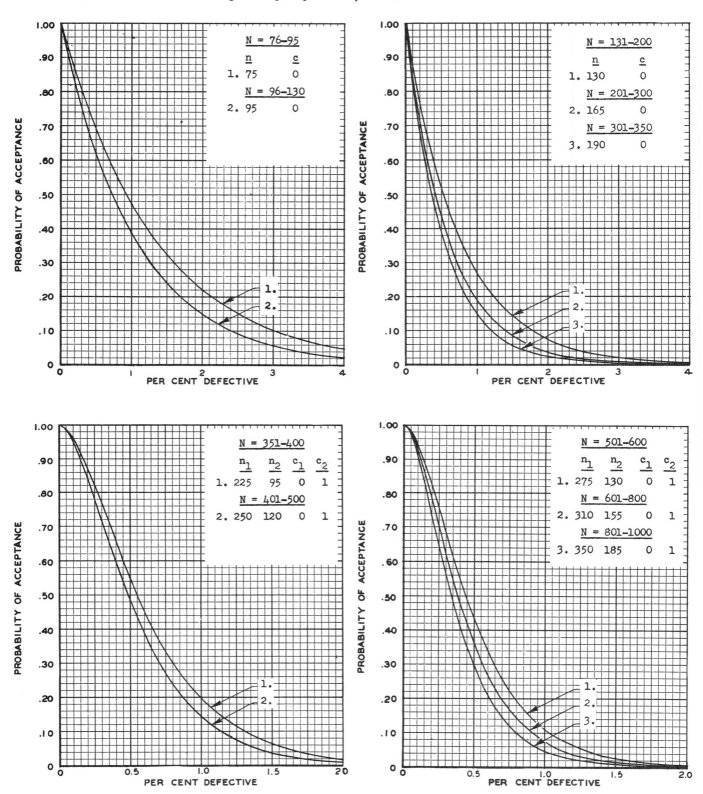

DOUBLE
SAMPLING
0.1%
A O Q L

Operating Characteristic Curves • Double Sampling Plans
Average Outgoing Quality Limit, AOQL = 0.1%

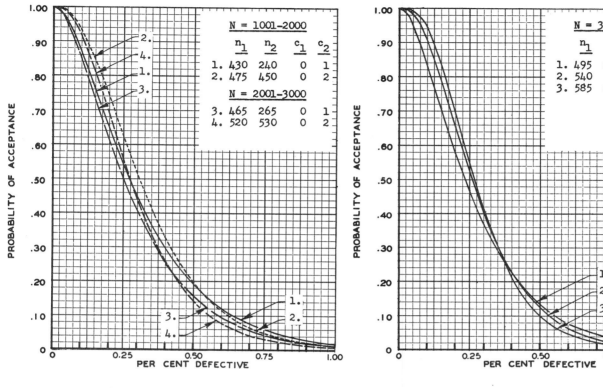

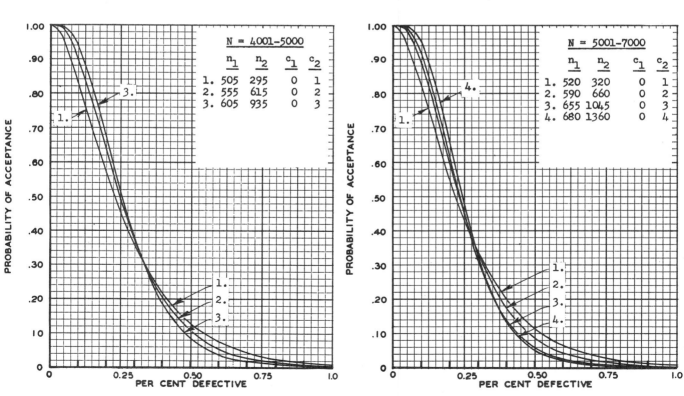

DOUBLE SAMPLING

0.1%
AOQL

Operating Characteristic Curves • Double Sampling Plans

Average Outgoing Quality Limit, AOQL = 0.1%

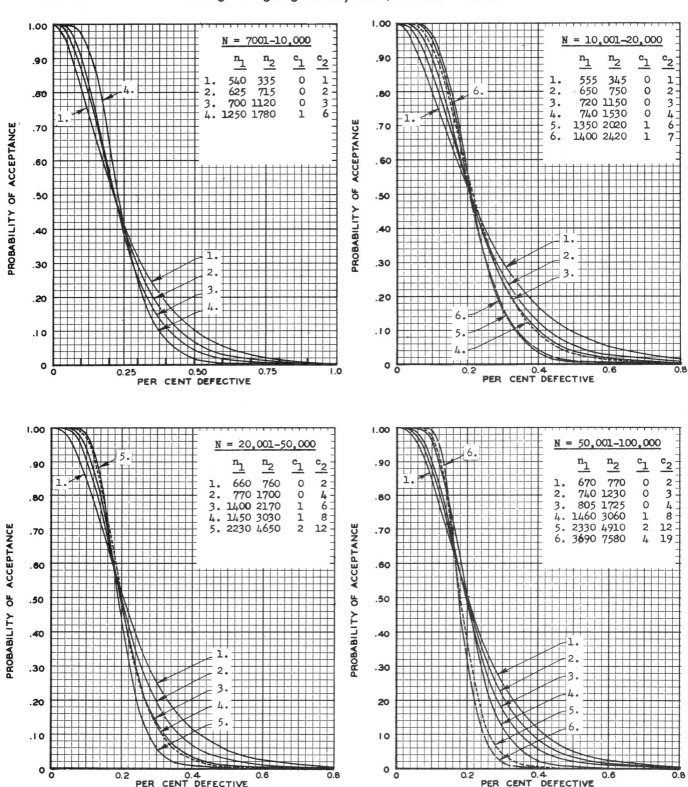

Operating Characteristic Curves • Double Sampling Plans

Average Outgoing Quality Limit, AOQL = 0.25%

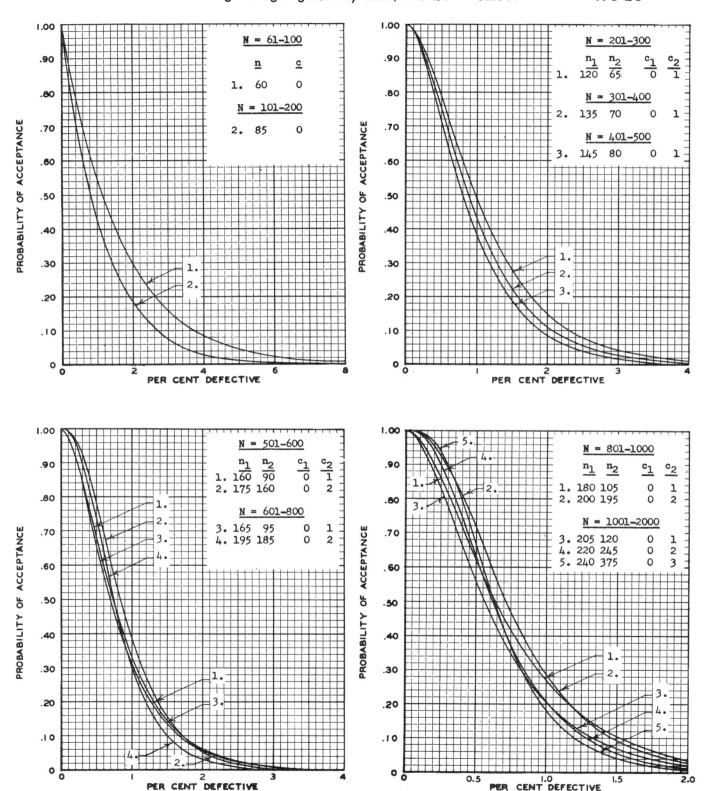

0.25%
AOQL

Operating Characteristic Curves • Double Sampling Plans

Average Outgoing Quality Limit, AOQL = 0.25%

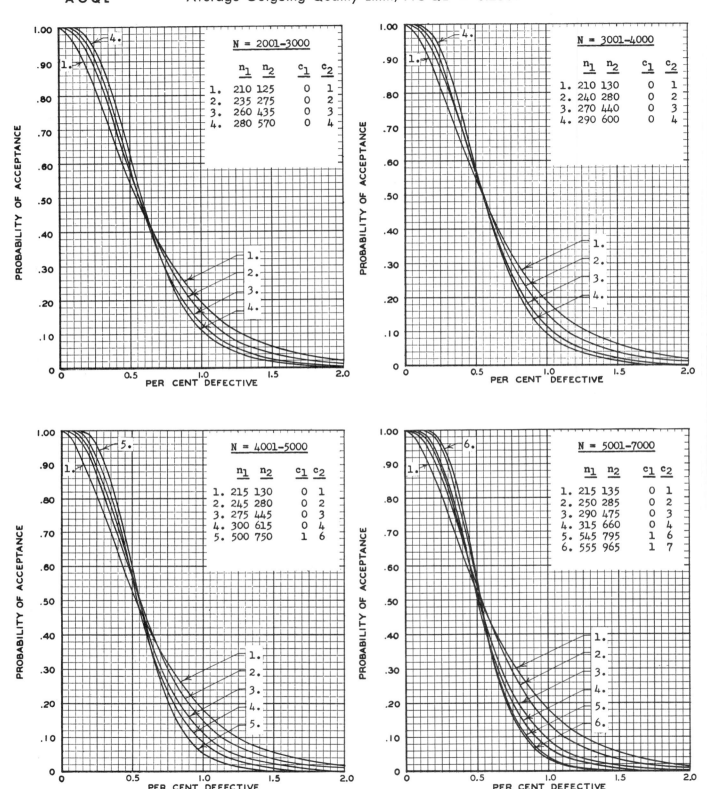

N = 2001-3000

	n_1	n_2	c_1	c_2
1.	210	125	0	1
2.	235	275	0	2
3.	260	435	0	3
4.	280	570	0	4

N = 3001-4000

	n_1	n_2	c_1	c_2
1.	210	130	0	1
2.	240	280	0	2
3.	270	440	0	3
4.	290	600	0	4

N = 4001-5000

	n_1	n_2	c_1	c_2
1.	215	130	0	1
2.	245	280	0	2
3.	275	445	0	3
4.	300	615	0	4
5.	500	750	1	6

N = 5001-7000

	n_1	n_2	c_1	c_2
1.	215	135	0	1
2.	250	285	0	2
3.	290	475	0	3
4.	315	660	0	4
5.	545	795	1	6
6.	555	965	1	7

Operating Characteristic Curves • Double Sampling Plans

Average Outgoing Quality Limit, AOQL = 0.25%

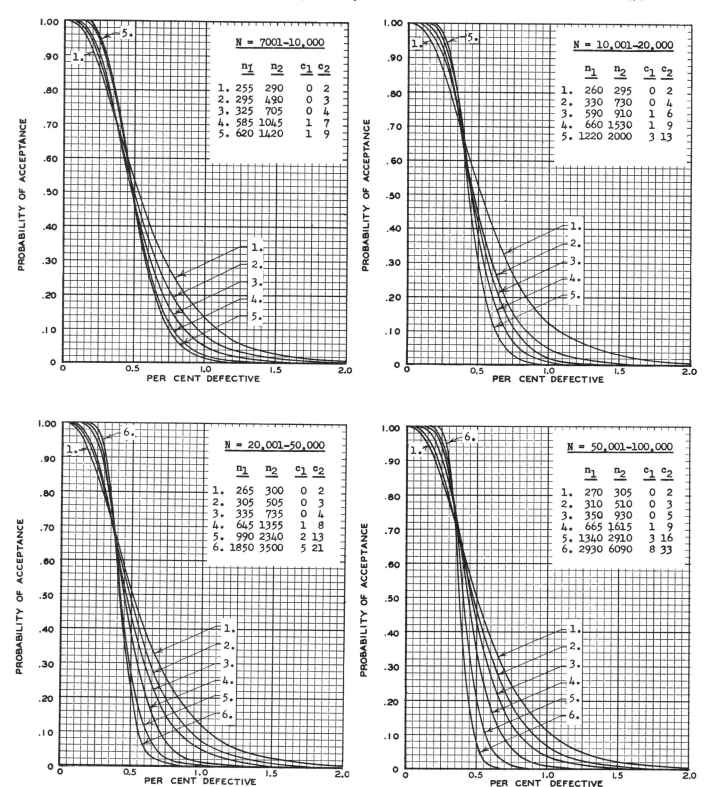

N = 7001-10,000

	n_1	n_2	c_1	c_2
1.	255	290	0	2
2.	295	490	0	3
3.	325	705	0	4
4.	585	1045	1	7
5.	620	1420	1	9

N = 10,001-20,000

	n_1	n_2	c_1	c_2
1.	260	295	0	2
2.	330	730	0	4
3.	590	910	1	6
4.	660	1530	1	9
5.	1220	2000	3	13

N = 20,001-50,000

	n_1	n_2	c_1	c_2
1.	265	300	0	2
2.	305	505	0	3
3.	335	735	0	4
4.	645	1355	1	8
5.	990	2340	2	13
6.	1850	3500	5	21

N = 50,001-100,000

	n_1	n_2	c_1	c_2
1.	270	305	0	2
2.	310	510	0	3
3.	350	930	0	5
4.	665	1615	1	9
5.	1340	2910	3	16
6.	2930	6090	8	33

PROBABILITY OF ACCEPTANCE

PER CENT DEFECTIVE

DOUBLE
SAMPLING

0.5%

AOQL

Operating Characteristic Curves • Double Sampling Plans
Average Outgoing Quality Limit, AOQL = 0.5%

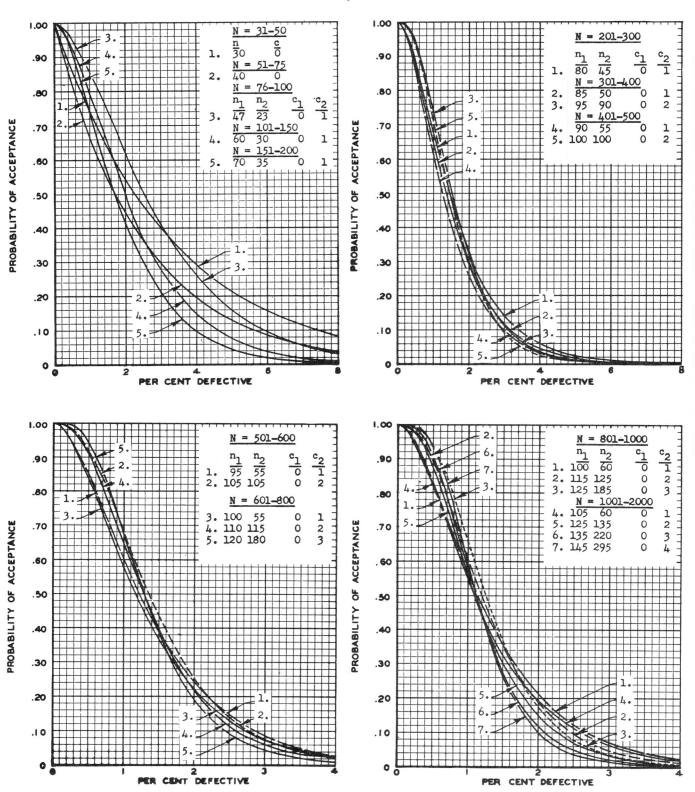

Operating Characteristic Curves • Double Sampling Plans
Average Outgoing Quality Limit, AOQL = 0.5%

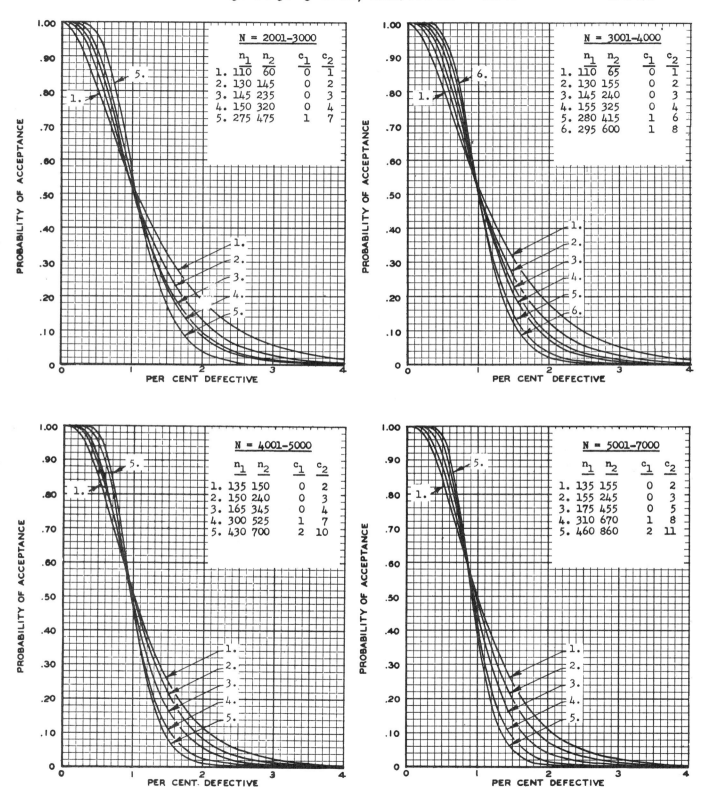

DOUBLE
SAMPLING

0.5%
AOQL

Operating Characteristic Curves • Double Sampling Plans
Average Outgoing Quality Limit, AOQL = 0.5%

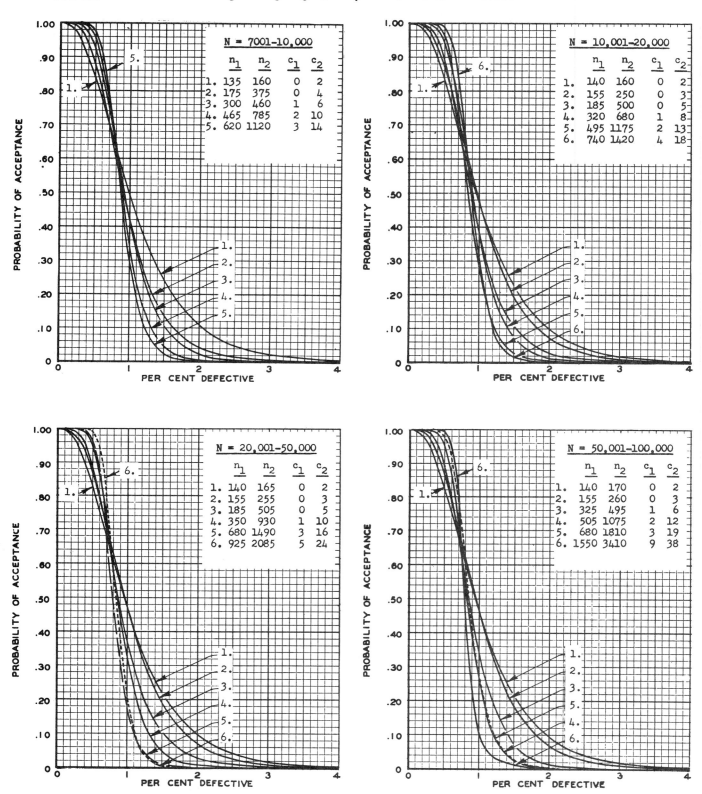

N = 7001–10,000

	n_1	n_2	c_1	c_2
1.	135	160	0	2
2.	175	375	0	4
3.	300	460	1	6
4.	465	785	2	10
5.	620	1120	3	14

N = 10,001–20,000

	n_1	n_2	c_1	c_2
1.	140	160	0	2
2.	155	250	0	3
3.	185	500	0	5
4.	320	680	1	8
5.	495	1175	2	13
6.	740	1420	4	18

N = 20,001–50,000

	n_1	n_2	c_1	c_2
1.	140	165	0	2
2.	155	255	0	3
3.	185	505	0	5
4.	350	930	1	10
5.	680	1490	3	16
6.	925	2085	5	24

N = 50,001–100,000

	n_1	n_2	c_1	c_2
1.	140	170	0	2
2.	155	260	0	3
3.	325	495	1	6
4.	505	1075	2	12
5.	680	1810	3	19
6.	1550	3410	9	38

PROBABILITY OF ACCEPTANCE

PER CENT DEFECTIVE

Operating Characteristic Curves • Double Sampling Plans

Average Outgoing Quality Limit, AOQL = 0.75%

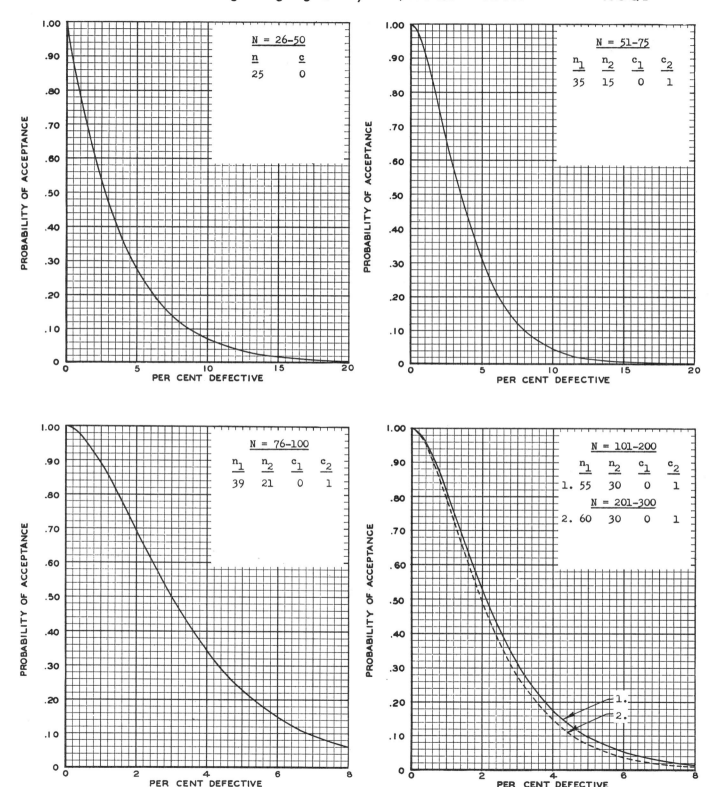

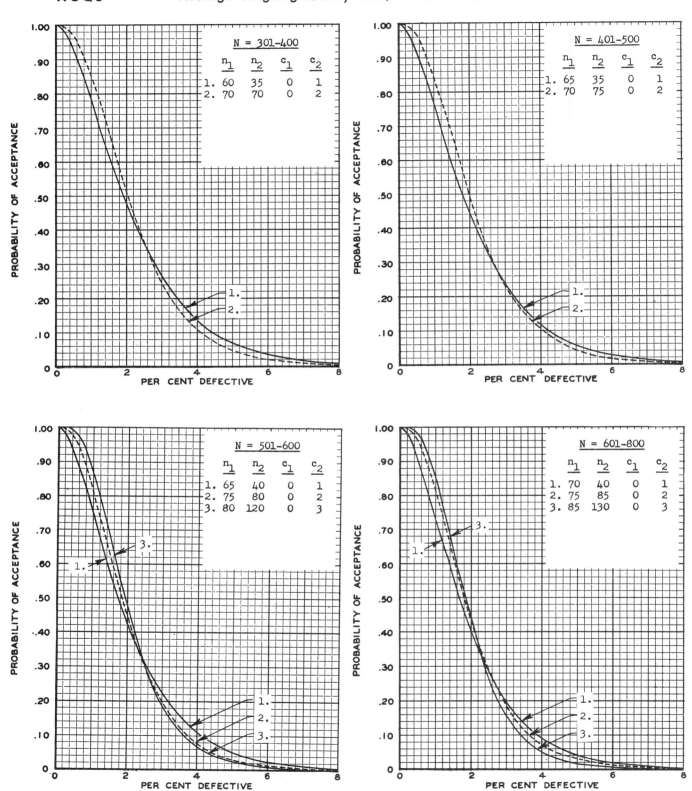

Operating Characteristic Curves • Double Sampling Plans

Average Outgoing Quality Limit, AOQL = 0.75%

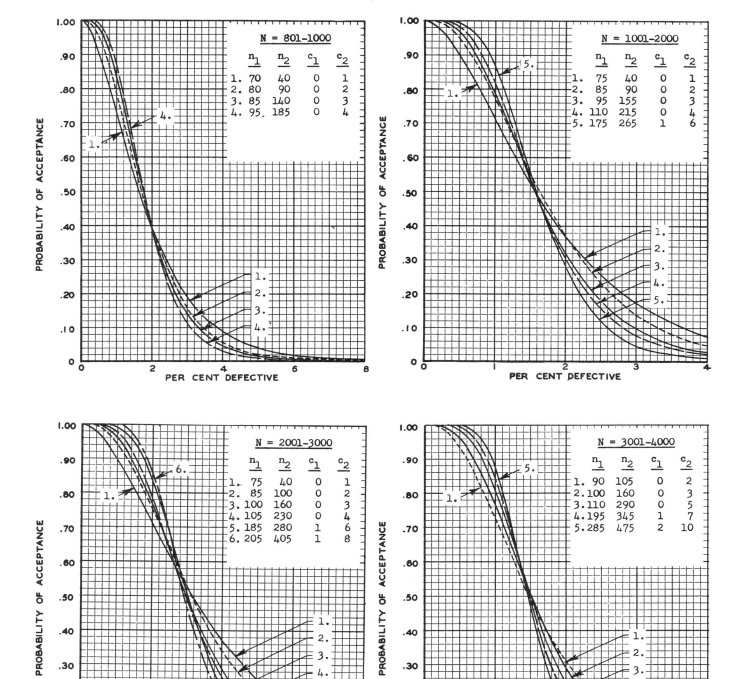

Operating Characteristic Curves • Double Sampling Plans

Average Outgoing Quality Limit, AOQL = 0.75%

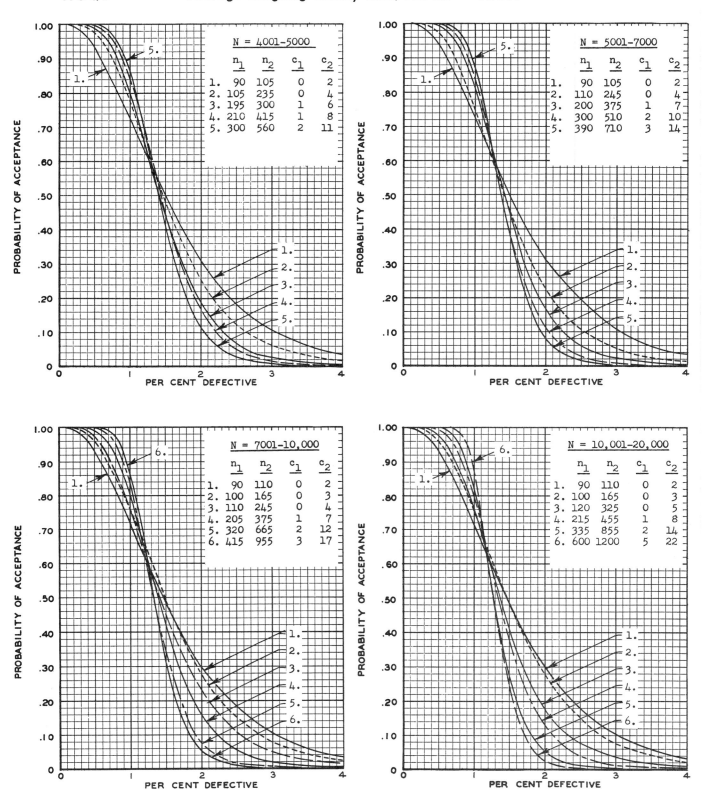

Operating Characteristic Curves • Double Sampling Plans

Average Outgoing Quality Limit, AOQL = 0.75%

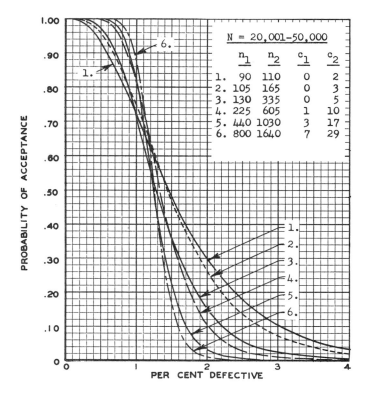

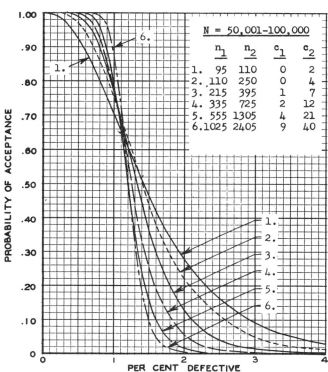

**DOUBLE
SAMPLING**

1.0%

AOQL

Operating Characteristic Curves • Double Sampling Plans
Average Outgoing Quality Limit, AOQL = 1.0%

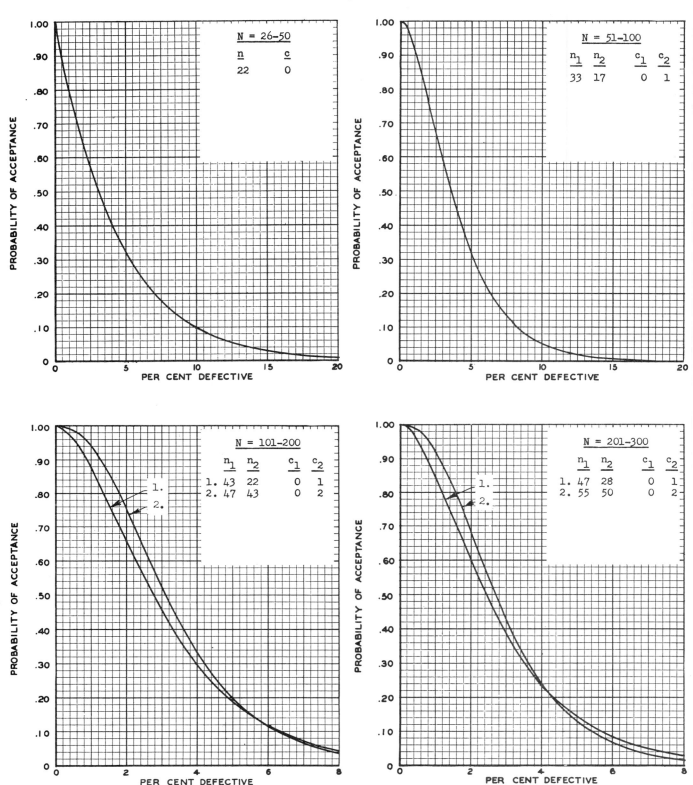

Operating Characteristic Curves • Double Sampling Plans
Average Outgoing Quality Limit, AOQL = 1.0%

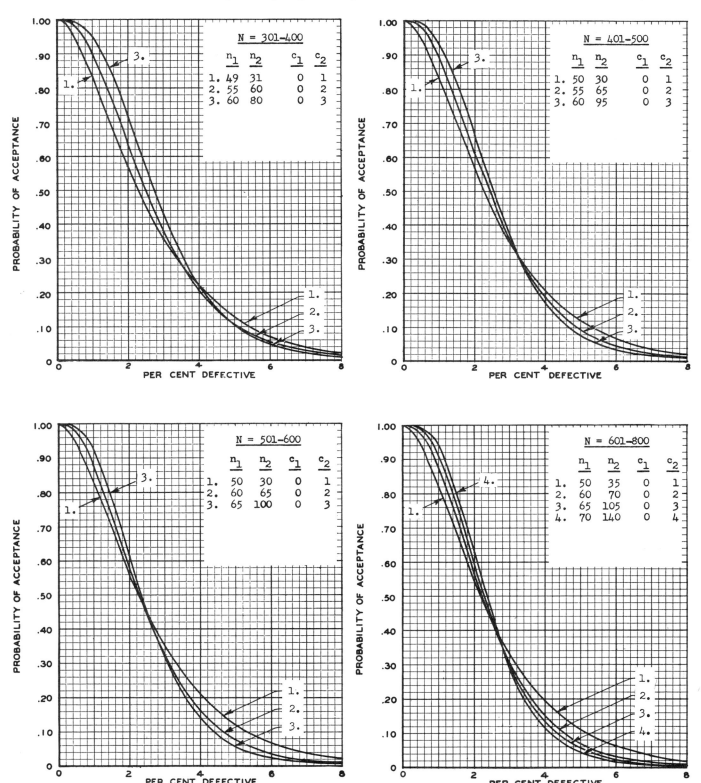

N = 301–400

	n_1	n_2	c_1	c_2
1.	49	31	0	1
2.	55	60	0	2
3.	60	80	0	3

N = 401–500

	n_1	n_2	c_1	c_2
1.	50	30	0	1
2.	55	65	0	2
3.	60	95	0	3

N = 501–600

	n_1	n_2	c_1	c_2
1.	50	30	0	1
2.	60	65	0	2
3.	65	100	0	3

N = 601–800

	n_1	n_2	c_1	c_2
1.	50	35	0	1
2.	60	70	0	2
3.	65	105	0	3
4.	70	140	0	4

Operating Characteristic Curves • Double Sampling Plans
Average Outgoing Quality Limit, AOQL = 1.0%

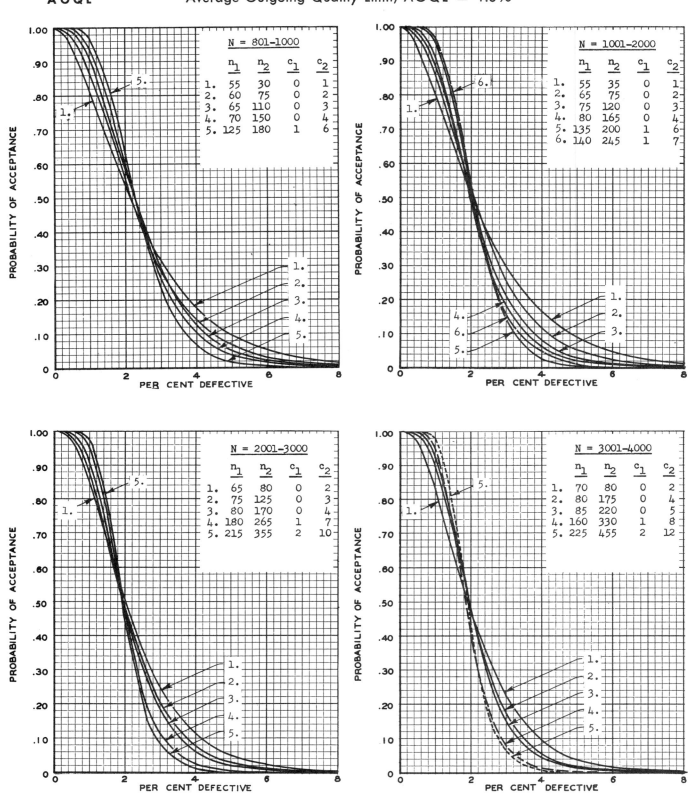

N = 801-1000			
n_1	n_2	c_1	c_2
1. 55	30	0	1
2. 60	75	0	2
3. 65	110	0	3
4. 70	150	0	4
5. 125	180	1	6

N = 1001-2000			
n_1	n_2	c_1	c_2
1. 55	35	0	1
2. 65	75	0	2
3. 75	120	0	3
4. 80	165	0	4
5. 135	200	1	6
6. 140	245	1	7

N = 2001-3000			
n_1	n_2	c_1	c_2
1. 65	80	0	2
2. 75	125	0	3
3. 80	170	0	4
4. 180	265	1	7
5. 215	355	2	10

N = 3001-4000			
n_1	n_2	c_1	c_2
1. 70	80	0	2
2. 80	175	0	4
3. 85	220	0	5
4. 160	330	1	8
5. 225	455	2	12

PROBABILITY OF ACCEPTANCE

PER CENT DEFECTIVE

Operating Characteristic Curves • Double Sampling Plans

Average Outgoing Quality Limit, AOQL = 1.0%

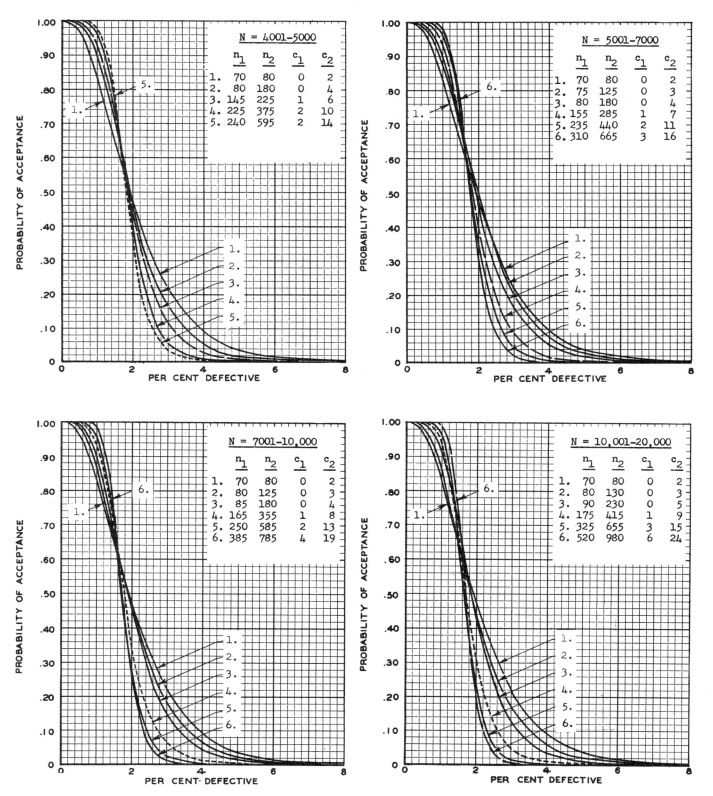

DOUBLE SAMPLING

1.0%

AOQL

Operating Characteristic Curves   •   Double Sampling Plans

Average Outgoing Quality Limit, AOQL = 1.0%

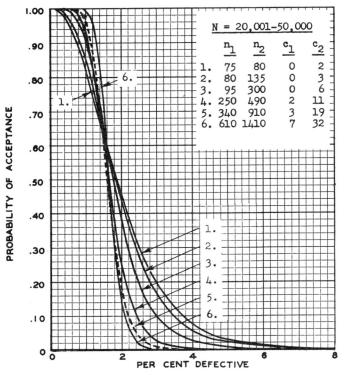

N = 20,001-50,000

	n_1	n_2	c_1	c_2
1.	75	80	0	2
2.	80	135	0	3
3.	95	300	0	6
4.	250	490	2	11
5.	340	910	3	19
6.	610	1410	7	32

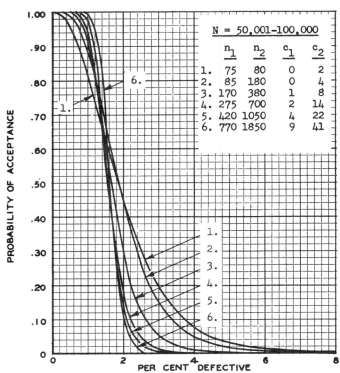

N = 50,001-100,000

	n_1	n_2	c_1	c_2
1.	75	80	0	2
2.	85	180	0	4
3.	170	380	1	8
4.	275	700	2	14
5.	420	1050	4	22
6.	770	1850	9	41

Operating Characteristic Curves • Double Sampling Plans

Average Outgoing Quality Limit, AOQL = 1.5%

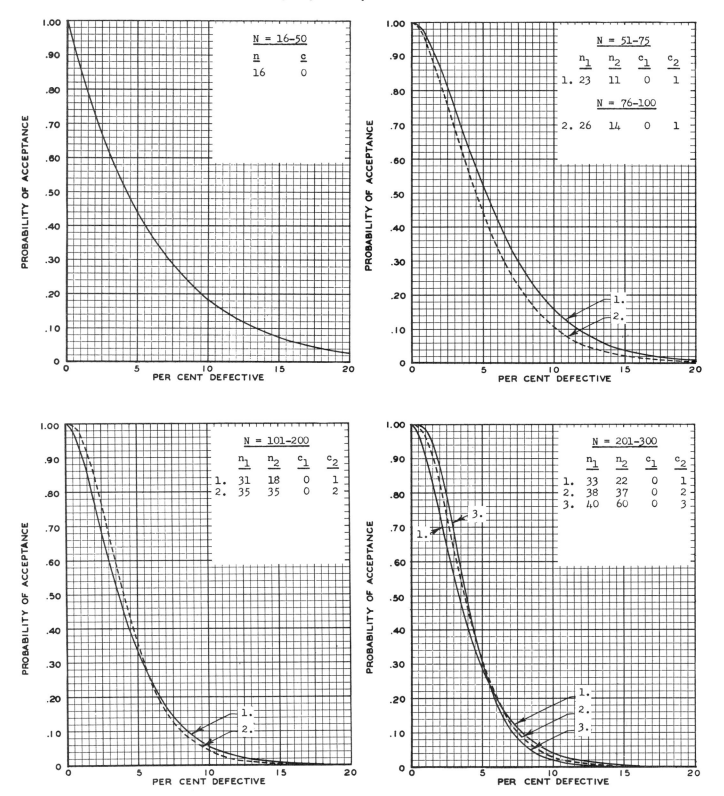

DOUBLE
SAMPLING

1.5%

AOQL

Operating Characteristic Curves • Double Sampling Plans
Average Outgoing Quality Limit, AOQL = 1.5%

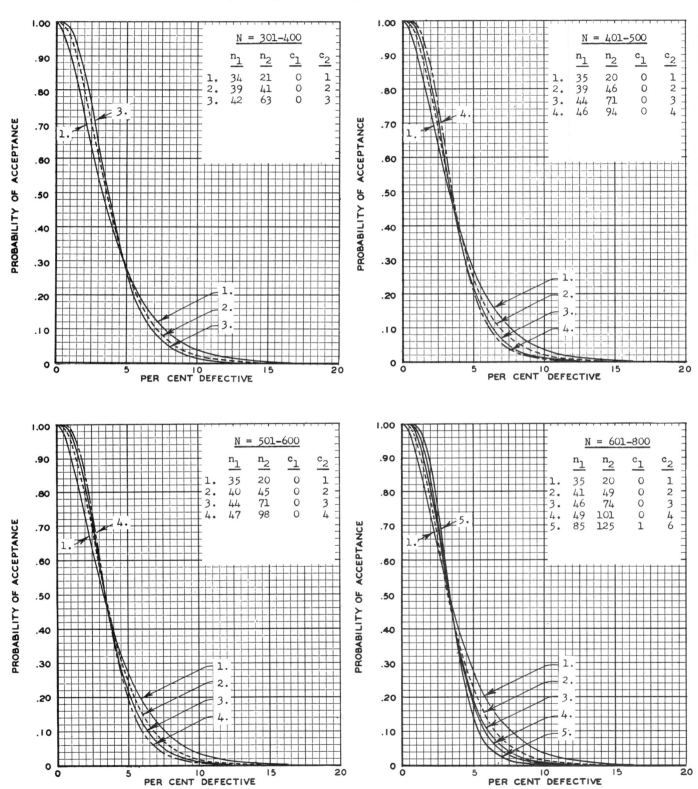

DOUBLE
SAMPLING
1.5%
A O Q L

Operating Characteristic Curves • Double Sampling Plans

Average Outgoing Quality Limit, AOQL = 1.5%

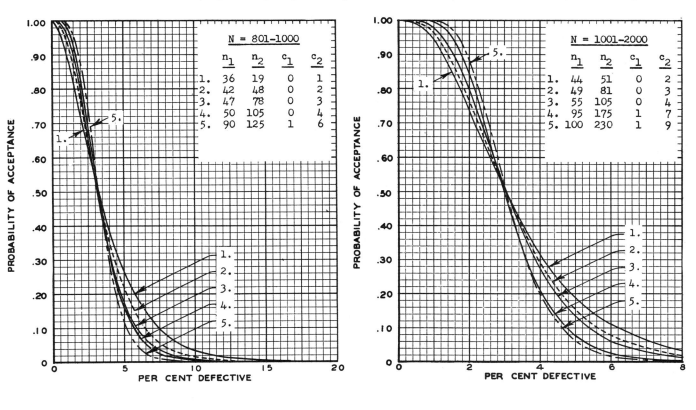

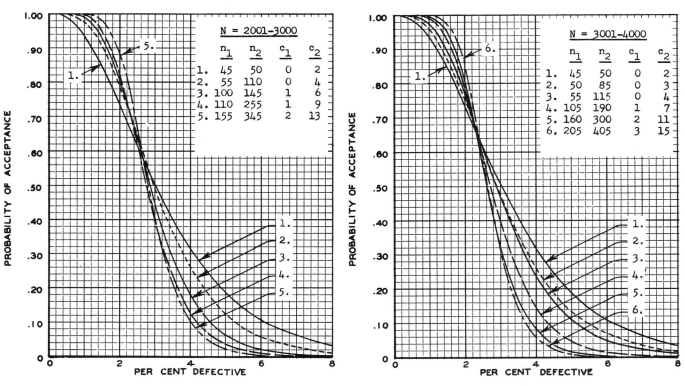

DOUBLE
SAMPLING

1.5%

AOQL

— 134 —

Appendix 2

Operating Characteristic Curves   •   Double Sampling Plans
Average Outgoing Quality Limit, AOQL = 1.5%

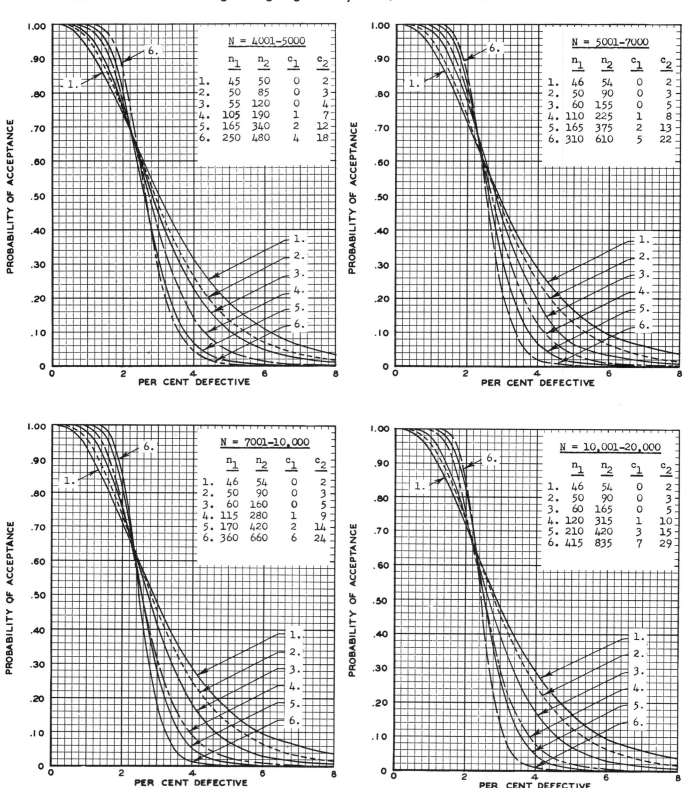

N = 4001-5000

	n_1	n_2	c_1	c_2
1.	45	50	0	2
2.	50	85	0	3
3.	55	120	0	4
4.	105	190	1	7
5.	165	340	2	12
6.	250	480	4	18

N = 5001-7000

	n_1	n_2	c_1	c_2
1.	46	54	0	2
2.	50	90	0	3
3.	60	155	0	5
4.	110	225	1	8
5.	165	375	2	13
6.	310	610	5	22

N = 7001-10,000

	n_1	n_2	c_1	c_2
1.	46	54	0	2
2.	50	90	0	3
3.	60	160	0	5
4.	115	280	1	9
5.	170	420	2	14
6.	360	660	6	24

N = 10,001-20,000

	n_1	n_2	c_1	c_2
1.	46	54	0	2
2.	50	90	0	3
3.	60	165	0	5
4.	120	315	1	10
5.	210	420	3	15
6.	415	835	7	29

PROBABILITY OF ACCEPTANCE

PER CENT DEFECTIVE

Operating Characteristic Curves • Double Sampling Plans
Average Outgoing Quality Limit, AOQL = 1.5%

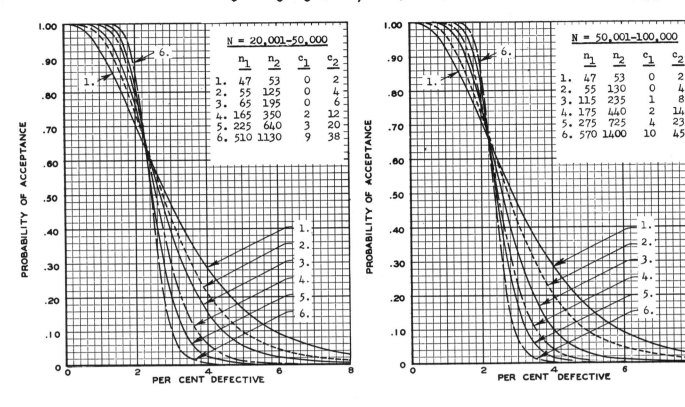

N = 20,001-50,000

	n_1	n_2	c_1	c_2
1.	47	53	0	2
2.	55	125	0	4
3.	65	195	0	6
4.	165	350	2	12
5.	225	640	3	20
6.	510	1130	9	38

N = 50,001-100,000

	n_1	n_2	c_1	c_2
1.	47	53	0	2
2.	55	130	0	4
3.	115	235	1	8
4.	175	440	2	14
5.	275	725	4	23
6.	570	1400	10	45

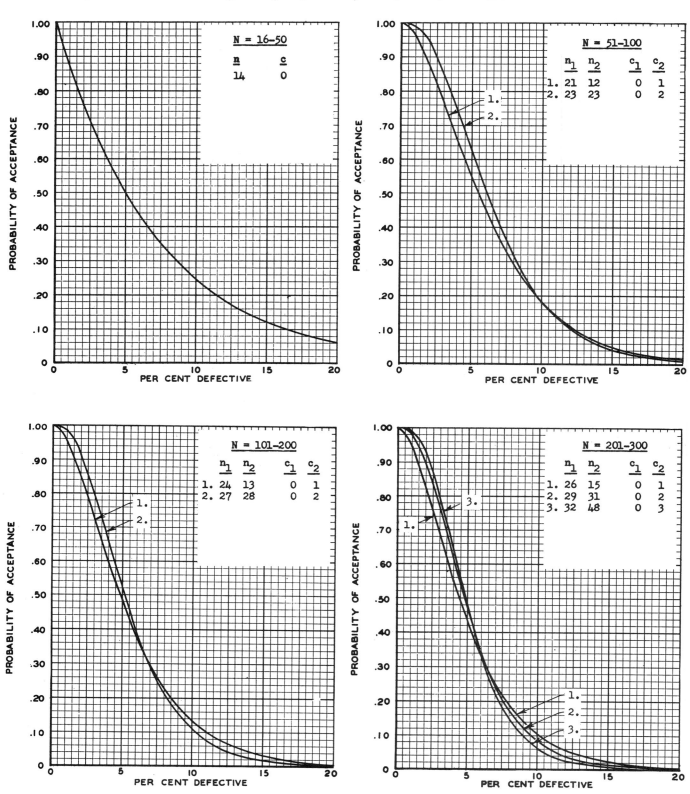

Operating Characteristic Curves • Double Sampling Plans
Average Outgoing Quality Limit, AOQL = 2.0%

N = 16-50

n	c
14	0

N = 51-100

	n_1	n_2	c_1	c_2
1.	21	12	0	1
2.	23	23	0	2

N = 101-200

	n_1	n_2	c_1	c_2
1.	24	13	0	1
2.	27	28	0	2

N = 201-300

	n_1	n_2	c_1	c_2
1.	26	15	0	1
2.	29	31	0	2
3.	32	48	0	3

DOUBLE
SAMPLING
2.0%
A O Q L

Operating Characteristic Curves • Double Sampling Plans

Average Outgoing Quality Limit, AOQL = 2.0%

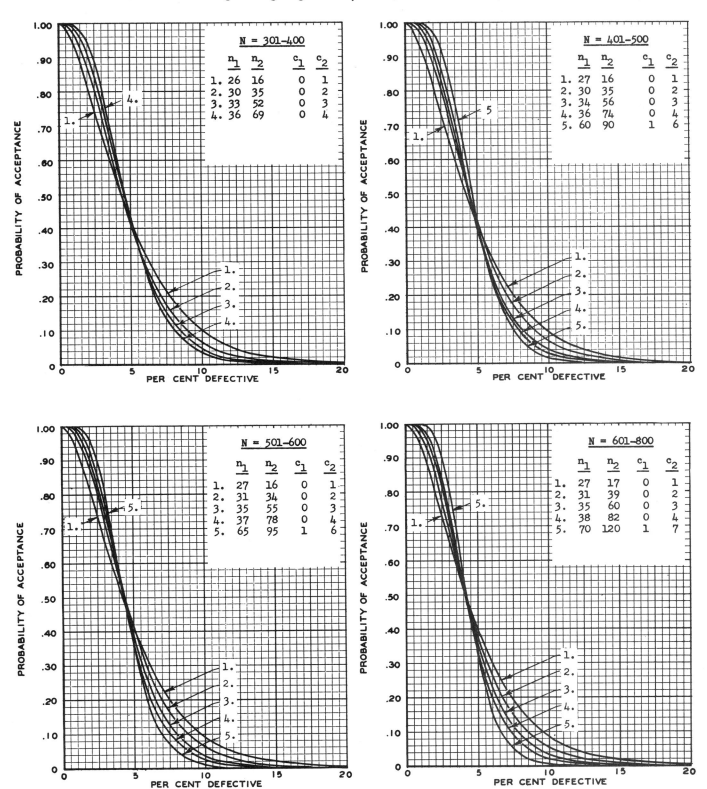

DOUBLE
SAMPLING

2.0%
AOQL

— 138 —

Appendix 2

Operating Characteristic Curves • Double Sampling Plans
Average Outgoing Quality Limit, AOQL = 2.0%

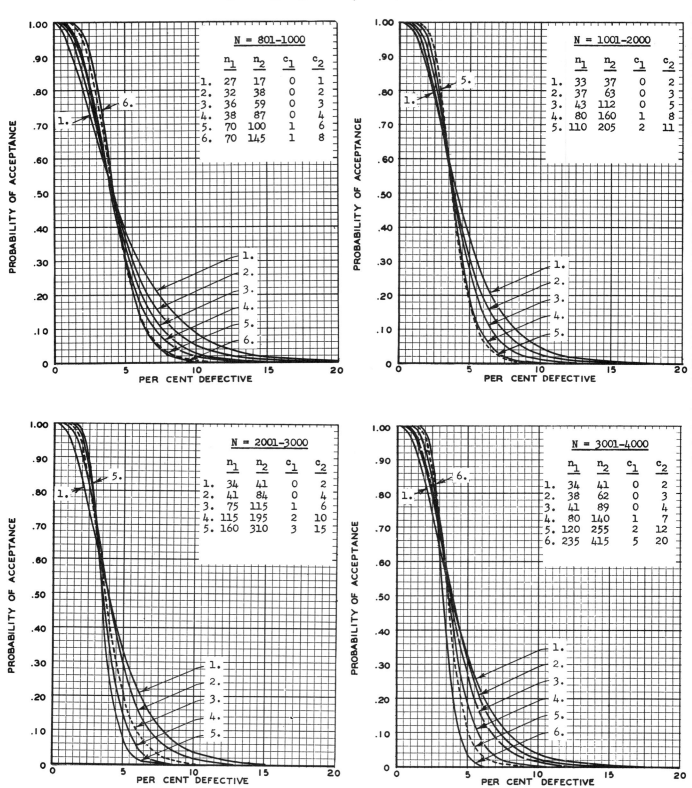

N = 801–1000

	n_1	n_2	c_1	c_2
1.	27	17	0	1
2.	32	38	0	2
3.	36	59	0	3
4.	38	87	0	4
5.	70	100	1	6
6.	70	145	1	8

N = 1001–2000

	n_1	n_2	c_1	c_2
1.	33	37	0	2
2.	37	63	0	3
3.	43	112	0	5
4.	80	160	1	8
5.	110	205	2	11

N = 2001–3000

	n_1	n_2	c_1	c_2
1.	34	41	0	2
2.	41	84	0	4
3.	75	115	1	6
4.	115	195	2	10
5.	160	310	3	15

N = 3001–4000

	n_1	n_2	c_1	c_2
1.	34	41	0	2
2.	38	62	0	3
3.	41	89	0	4
4.	80	140	1	7
5.	120	255	2	12
6.	235	415	5	20

PROBABILITY OF ACCEPTANCE

PER CENT DEFECTIVE

Operating Characteristic Curves • Double Sampling Plans
Average Outgoing Quality Limit, AOQL = 2.0%

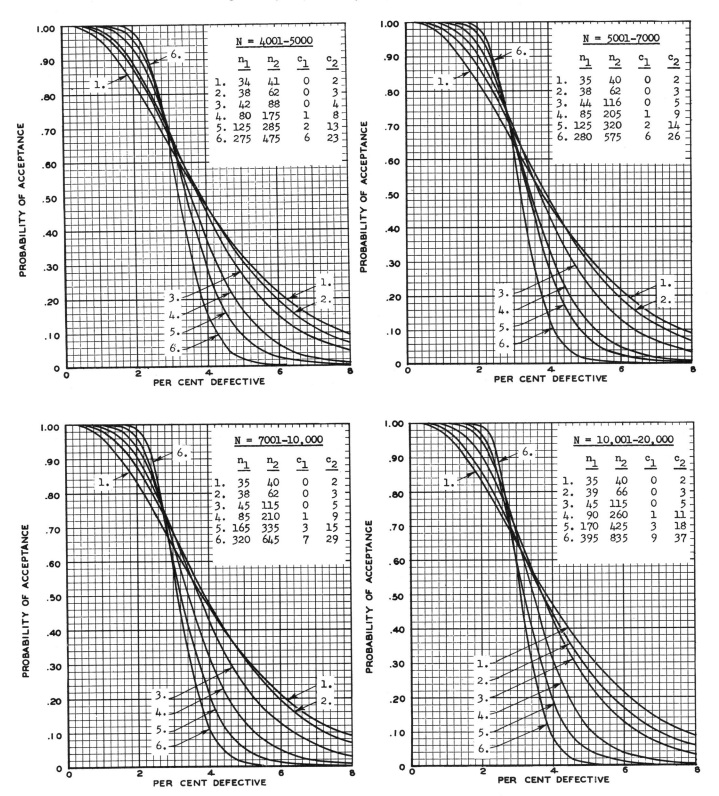

N = 4001-5000

	n_1	n_2	c_1	c_2
1.	34	41	0	2
2.	38	62	0	3
3.	42	88	0	4
4.	80	175	1	8
5.	125	285	2	13
6.	275	475	6	23

N = 5001-7000

	n_1	n_2	c_1	c_2
1.	35	40	0	2
2.	38	62	0	3
3.	44	116	0	5
4.	85	205	1	9
5.	125	320	2	14
6.	280	575	6	26

N = 7001-10,000

	n_1	n_2	c_1	c_2
1.	35	40	0	2
2.	38	62	0	3
3.	45	115	0	5
4.	85	210	1	9
5.	165	335	3	15
6.	320	645	7	29

N = 10,001-20,000

	n_1	n_2	c_1	c_2
1.	35	40	0	2
2.	39	66	0	3
3.	45	115	0	5
4.	90	260	1	11
5.	170	425	3	18
6.	395	835	9	37

PROBABILITY OF ACCEPTANCE

PER CENT DEFECTIVE

DOUBLE
SAMPLING
2.0%
AOQL

Operating Characteristic Curves   •   Double Sampling Plans
Average Outgoing Quality Limit, AOQL = 2.0%

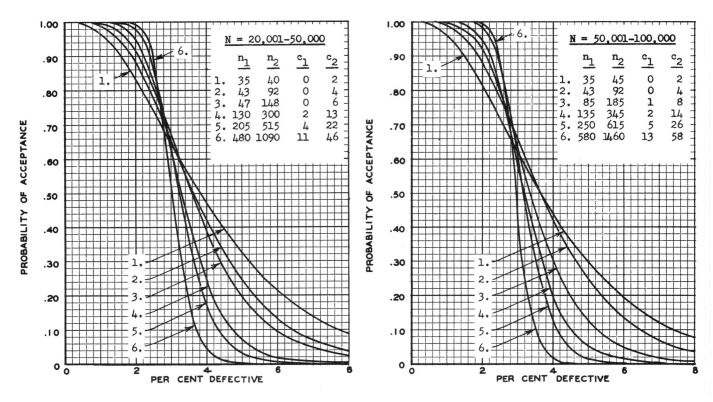

$N = 20,001-50,000$

	n_1	n_2	c_1	c_2
1.	35	40	0	2
2.	43	92	0	4
3.	47	148	0	6
4.	130	300	2	13
5.	205	515	4	22
6.	480	1090	11	46

$N = 50,001-100,000$

	n_1	n_2	c_1	c_2
1.	35	45	0	2
2.	43	92	0	4
3.	85	185	1	8
4.	135	345	2	14
5.	250	615	5	26
6.	580	1460	13	58

Operating Characteristic Curves • Double Sampling Plans
Average Outgoing Quality Limit, AOQL = 2.5%

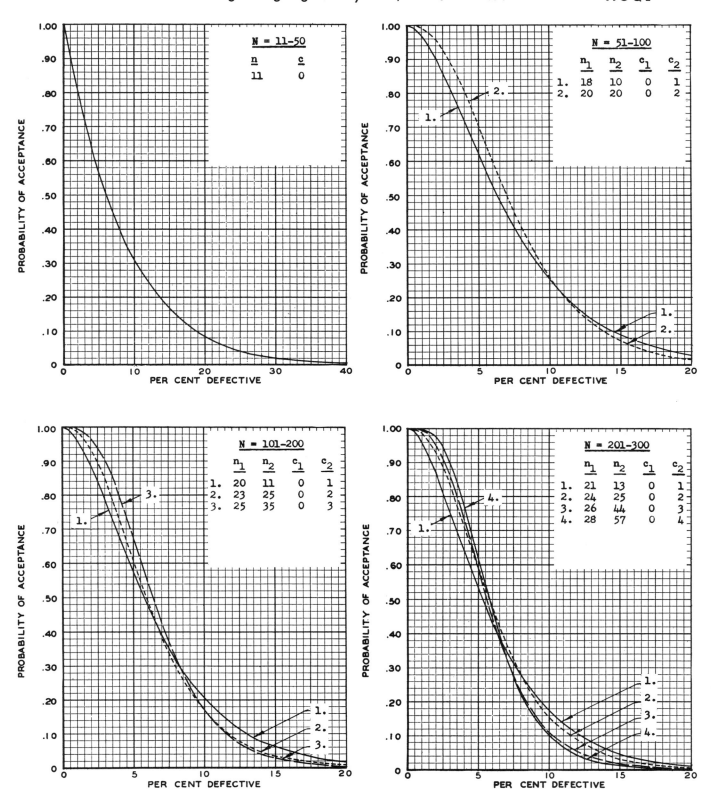

N = 11-50

n	c
11	0

N = 51-100

	n₁	n₂	c₁	c₂
1.	18	10	0	1
2.	20	20	0	2

N = 101-200

	n₁	n₂	c₁	c₂
1.	20	11	0	1
2.	23	25	0	2
3.	25	35	0	3

N = 201-300

	n₁	n₂	c₁	c₂
1.	21	13	0	1
2.	24	25	0	2
3.	26	44	0	3
4.	28	57	0	4

2.5%
AOQL

Operating Characteristic Curves • Double Sampling Plans

Average Outgoing Quality Limit, AOQL = 2.5%

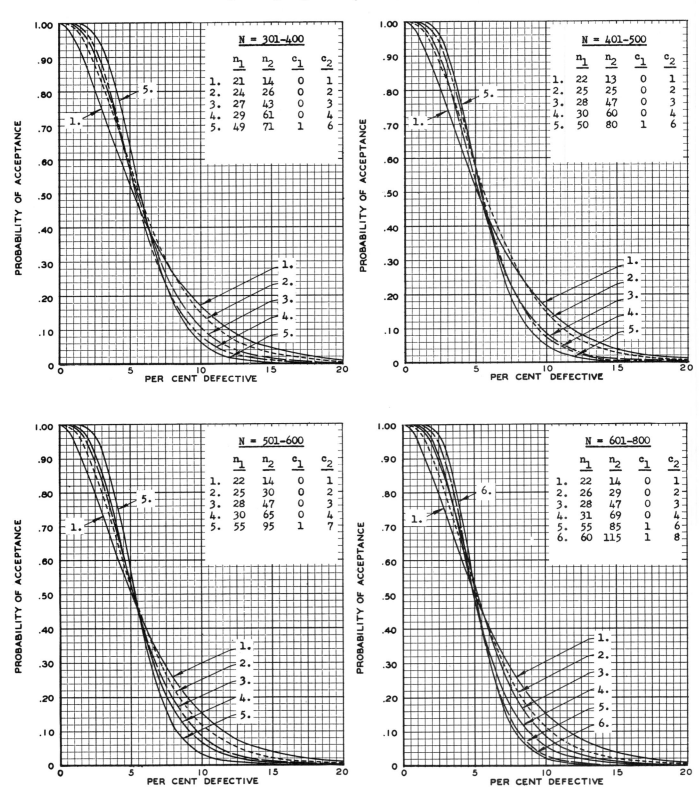

N = 301-400

	n_1	n_2	c_1	c_2
1.	21	14	0	1
2.	24	26	0	2
3.	27	43	0	3
4.	29	61	0	4
5.	49	71	1	6

N = 401-500

	n_1	n_2	c_1	c_2
1.	22	13	0	1
2.	25	25	0	2
3.	28	47	0	3
4.	30	60	0	4
5.	50	80	1	6

N = 501-600

	n_1	n_2	c_1	c_2
1.	22	14	0	1
2.	25	30	0	2
3.	28	47	0	3
4.	30	65	0	4
5.	55	95	1	7

N = 601-800

	n_1	n_2	c_1	c_2
1.	22	14	0	1
2.	26	29	0	2
3.	28	47	0	3
4.	31	69	0	4
5.	55	85	1	6
6.	60	115	1	8

PROBABILITY OF ACCEPTANCE

PER CENT DEFECTIVE

Operating Characteristic Curves • Double Sampling Plans
Average Outgoing Quality Limit, AOQL = 2.5%

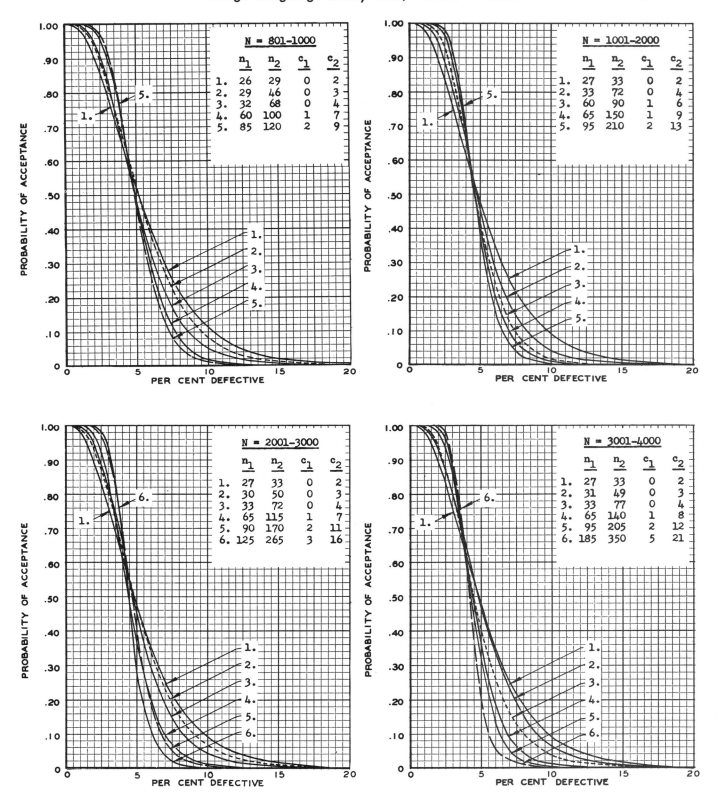

N = 801-1000

	n_1	n_2	c_1	c_2
1.	26	29	0	2
2.	29	46	0	3
3.	32	68	0	4
4.	60	100	1	7
5.	85	120	2	9

N = 1001-2000

	n_1	n_2	c_1	c_2
1.	27	33	0	2
2.	33	72	0	4
3.	60	90	1	6
4.	65	150	1	9
5.	95	210	2	13

N = 2001-3000

	n_1	n_2	c_1	c_2
1.	27	33	0	2
2.	30	50	0	3
3.	33	72	0	4
4.	65	115	1	7
5.	90	170	2	11
6.	125	265	3	16

N = 3001-4000

	n_1	n_2	c_1	c_2
1.	27	33	0	2
2.	31	49	0	3
3.	33	77	0	4
4.	65	140	1	8
5.	95	205	2	12
6.	185	350	5	21

PROBABILITY OF ACCEPTANCE

PER CENT DEFECTIVE

2.5% AOQL

Operating Characteristic Curves • Double Sampling Plans

Average Outgoing Quality Limit, AOQL = 2.5%

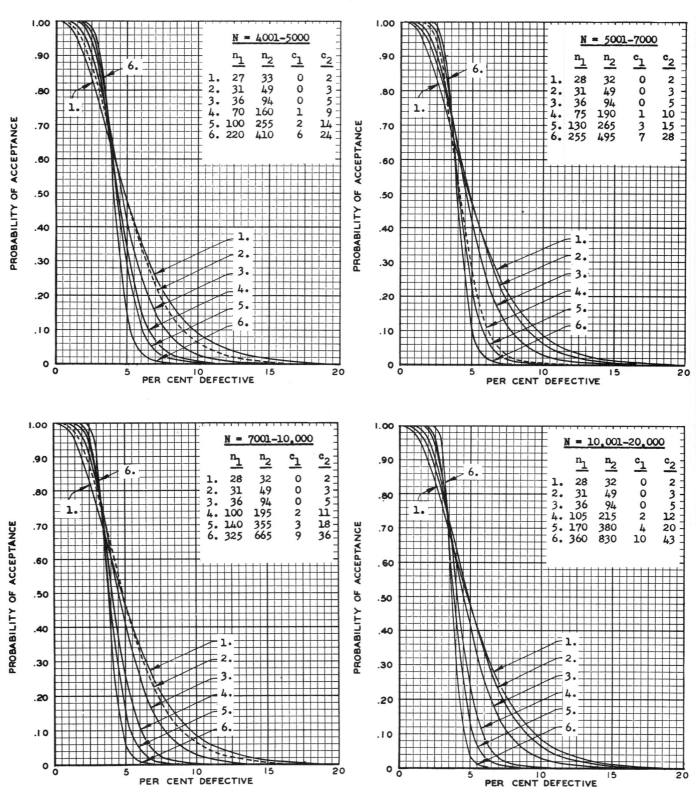

N = 4001-5000

	n_1	n_2	c_1	c_2
1.	27	33	0	2
2.	31	49	0	3
3.	36	94	0	5
4.	70	160	1	9
5.	100	255	2	14
6.	220	410	6	24

N = 5001-7000

	n_1	n_2	c_1	c_2
1.	28	32	0	2
2.	31	49	0	3
3.	36	94	0	5
4.	75	190	1	10
5.	130	265	3	15
6.	255	495	7	28

N = 7001-10,000

	n_1	n_2	c_1	c_2
1.	28	32	0	2
2.	31	49	0	3
3.	36	94	0	5
4.	100	195	2	11
5.	140	355	3	18
6.	325	665	9	36

N = 10,001-20,000

	n_1	n_2	c_1	c_2
1.	28	32	0	2
2.	31	49	0	3
3.	36	94	0	5
4.	105	215	2	12
5.	170	380	4	20
6.	360	830	10	43

DOUBLE
SAMPLING
2.5%
A O Q L

Operating Characteristic Curves • Double Sampling Plans
Average Outgoing Quality Limit, AOQL = 2.5%

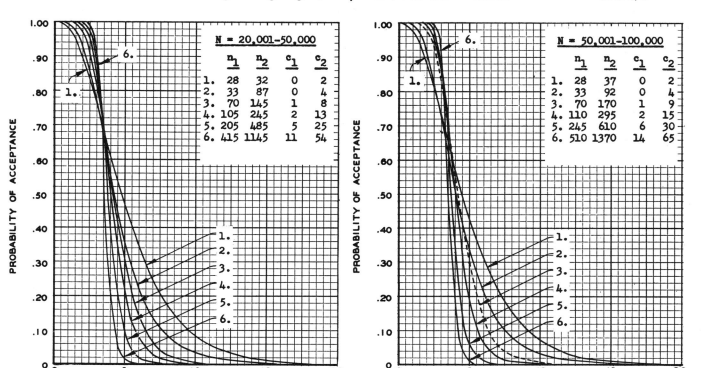

N = 20,001-50,000

	n_1	n_2	c_1	c_2
1.	28	32	0	2
2.	33	87	0	4
3.	70	145	1	8
4.	105	245	2	13
5.	205	485	5	25
6.	415	1145	11	54

N = 50,001-100,000

	n_1	n_2	c_1	c_2
1.	28	37	0	2
2.	33	92	0	4
3.	70	170	1	9
4.	110	295	2	15
5.	245	610	6	30
6.	510	1370	14	65

DOUBLE
SAMPLING

3.0%
AOQL

Operating Characteristic Curves • Double Sampling Plans
Average Outgoing Quality Limit, AOQL = 3.0%

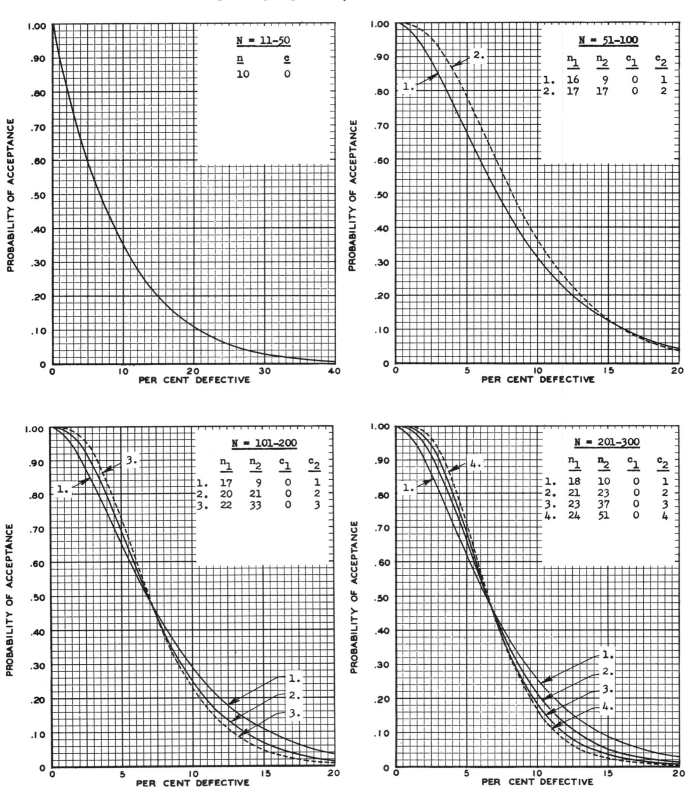

N = 11-50

n	c
10	0

N = 51-100

	n_1	n_2	c_1	c_2
1.	16	9	0	1
2.	17	17	0	2

N = 101-200

	n_1	n_2	c_1	c_2
1.	17	9	0	1
2.	20	21	0	2
3.	22	33	0	3

N = 201-300

	n_1	n_2	c_1	c_2
1.	18	10	0	1
2.	21	23	0	2
3.	23	37	0	3
4.	24	51	0	4

Operating Characteristic Curves • Double Sampling Plans

Average Outgoing Quality Limit, AOQL = 3.0%

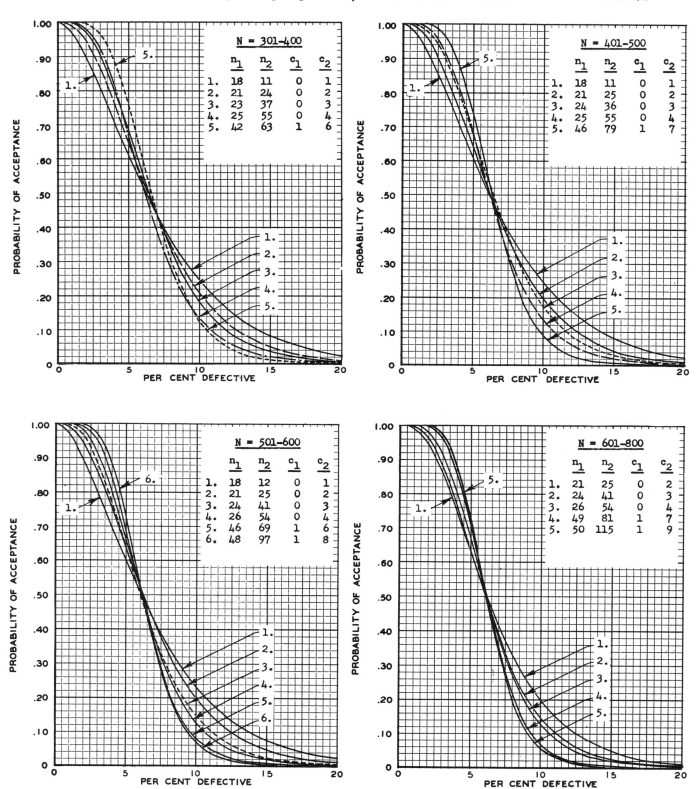

N = 301–400

	n_1	n_2	c_1	c_2
1.	18	11	0	1
2.	21	24	0	2
3.	23	37	0	3
4.	25	55	0	4
5.	42	63	1	6

N = 401–500

	n_1	n_2	c_1	c_2
1.	18	11	0	1
2.	21	25	0	2
3.	24	36	0	3
4.	25	55	0	4
5.	46	79	1	7

N = 501–600

	n_1	n_2	c_1	c_2
1.	18	12	0	1
2.	21	25	0	2
3.	24	41	0	3
4.	26	54	0	4
5.	46	69	1	6
6.	48	97	1	8

N = 601–800

	n_1	n_2	c_1	c_2
1.	21	25	0	2
2.	24	41	0	3
3.	26	54	0	4
4.	49	81	1	7
5.	50	115	1	9

PROBABILITY OF ACCEPTANCE

PER CENT DEFECTIVE

DOUBLE SAMPLING

3.0%
A O Q L

Operating Characteristic Curves • Double Sampling Plans
Average Outgoing Quality Limit, AOQL = 3.0%

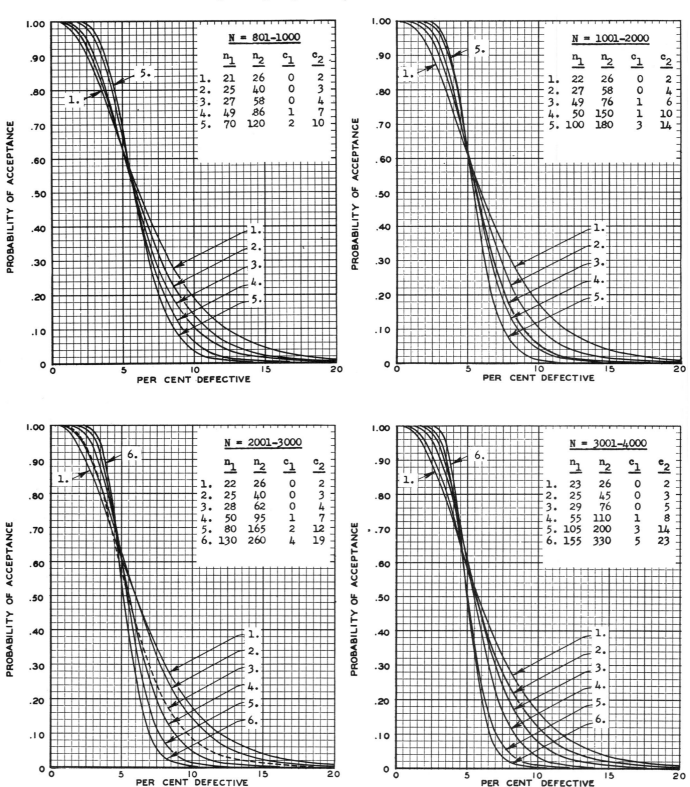

Operating Characteristic Curves • Double Sampling Plans
Average Outgoing Quality Limit, AOQL = 3.0%

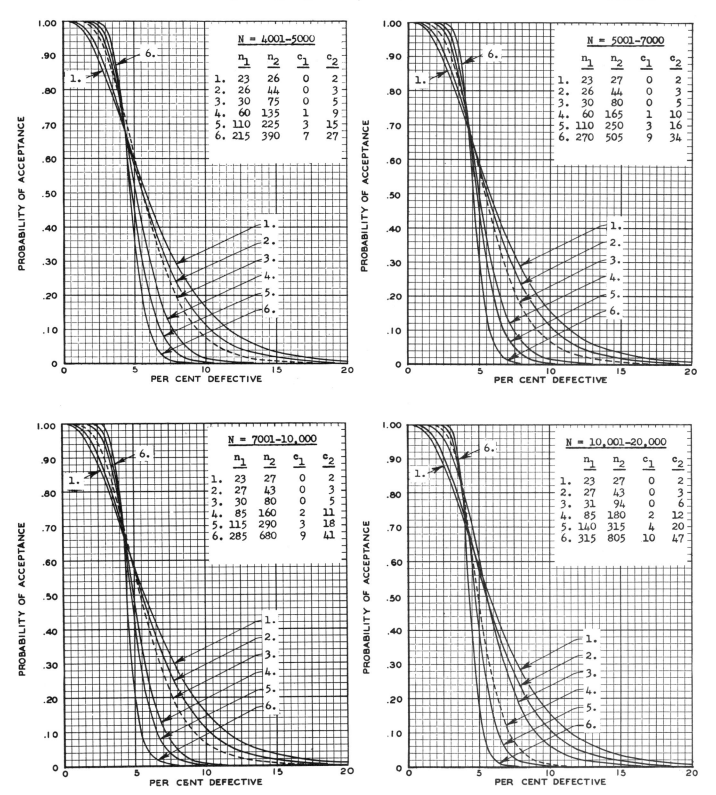

N = 4001-5000

	n_1	n_2	c_1	c_2
1.	23	26	0	2
2.	26	44	0	3
3.	30	75	0	5
4.	60	135	1	9
5.	110	225	3	15
6.	215	390	7	27

N = 5001-7000

	n_1	n_2	c_1	c_2
1.	23	27	0	2
2.	26	44	0	3
3.	30	80	0	5
4.	60	165	1	10
5.	110	250	3	16
6.	270	505	9	34

N = 7001-10,000

	n_1	n_2	c_1	c_2
1.	23	27	0	2
2.	27	43	0	3
3.	30	80	0	5
4.	85	160	2	11
5.	115	290	3	18
6.	285	680	9	41

N = 10,001-20,000

	n_1	n_2	c_1	c_2
1.	23	27	0	2
2.	27	43	0	3
3.	31	94	0	6
4.	85	180	2	12
5.	140	315	4	20
6.	315	805	10	47

DOUBLE
SAMPLING
3.0%
AOQL

Operating Characteristic Curves • Double Sampling Plans
Average Outgoing Quality Limit, AOQL = 3.0%

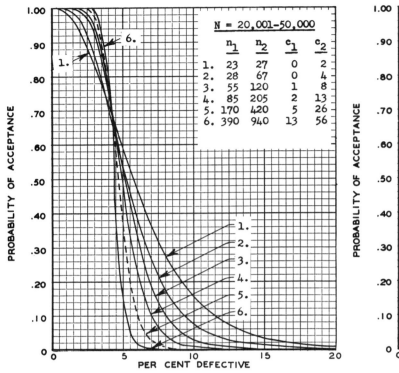

$N = 20,001-50,000$

	n_1	n_2	c_1	c_2
1.	23	27	0	2
2.	28	67	0	4
3.	55	120	1	8
4.	85	205	2	13
5.	170	420	5	26
6.	390	940	13	56

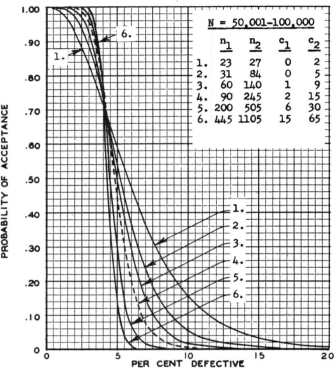

$N = 50,001-100,000$

	n_1	n_2	c_1	c_2
1.	23	27	0	2
2.	31	84	0	5
3.	60	140	1	9
4.	90	245	2	15
5.	200	505	6	30
6.	445	1105	15	65

Operating Characteristic Curves • Double Sampling Plans
Average Outgoing Quality Limit, AOQL = 4.0%

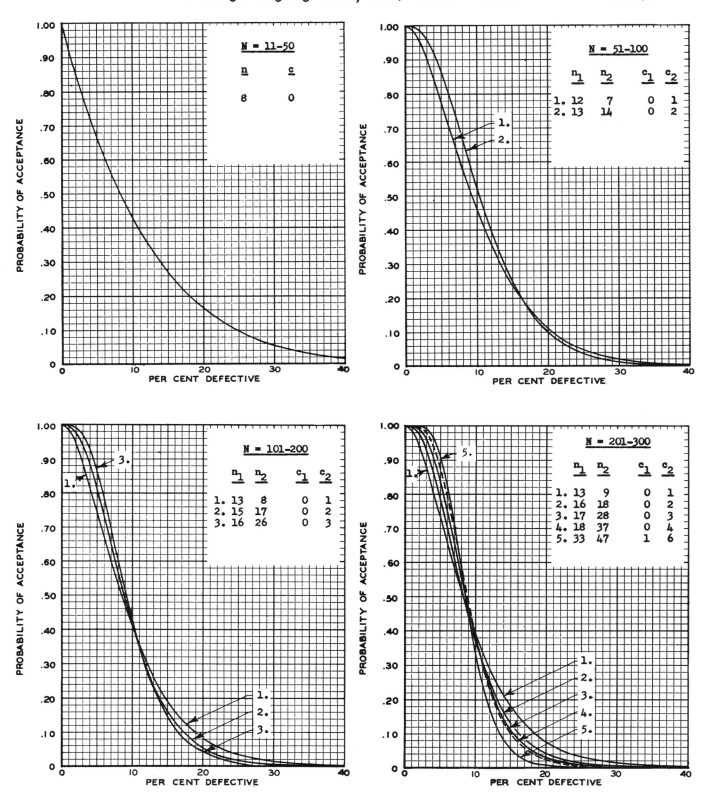

N = 11-50

n	c
8	0

N = 51-100

	n_1	n_2	c_1	c_2
1.	12	7	0	1
2.	13	14	0	2

N = 101-200

	n_1	n_2	c_1	c_2
1.	13	8	0	1
2.	15	17	0	2
3.	16	26	0	3

N = 201-300

	n_1	n_2	c_1	c_2
1.	13	9	0	1
2.	16	18	0	2
3.	17	28	0	3
4.	18	37	0	4
5.	33	47	1	6

PROBABILITY OF ACCEPTANCE

PER CENT DEFECTIVE

DOUBLE
SAMPLING

4.0%
AOQL

Operating Characteristic Curves • Double Sampling Plans
Average Outgoing Quality Limit, AOQL = 4.0%

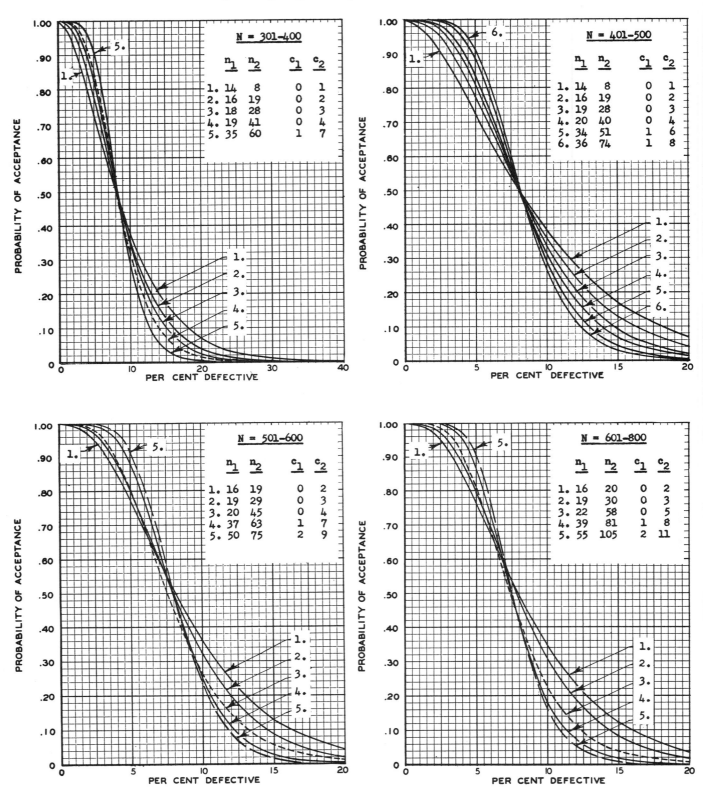

N = 301-400

	n₁	n₂	c₁	c₂
1.	14	8	0	1
2.	16	19	0	2
3.	18	28	0	3
4.	19	41	0	4
5.	35	60	1	7

N = 401-500

	n₁	n₂	c₁	c₂
1.	14	8	0	1
2.	16	19	0	2
3.	19	28	0	3
4.	20	40	0	4
5.	34	51	1	6
6.	36	74	1	8

N = 501-600

	n₁	n₂	c₁	c₂
1.	16	19	0	2
2.	19	29	0	3
3.	20	45	0	4
4.	37	63	1	7
5.	50	75	2	9

N = 601-800

	n₁	n₂	c₁	c₂
1.	16	20	0	2
2.	19	30	0	3
3.	22	58	0	5
4.	39	81	1	8
5.	55	105	2	11

PROBABILITY OF ACCEPTANCE

PER CENT DEFECTIVE

DOUBLE
SAMPLING
4.0%
A O Q L

Operating Characteristic Curves • Double Sampling Plans
Average Outgoing Quality Limit, AOQL = 4.0%

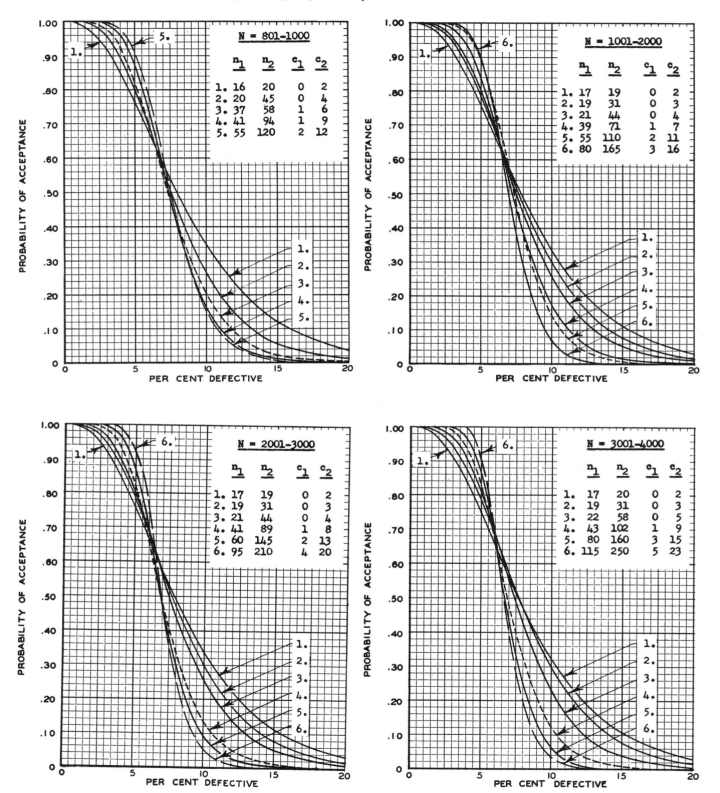

DOUBLE
SAMPLING

4.0%

AOQL

Operating Characteristic Curves • Double Sampling Plans

Average Outgoing Quality Limit, AOQL = 4.0%

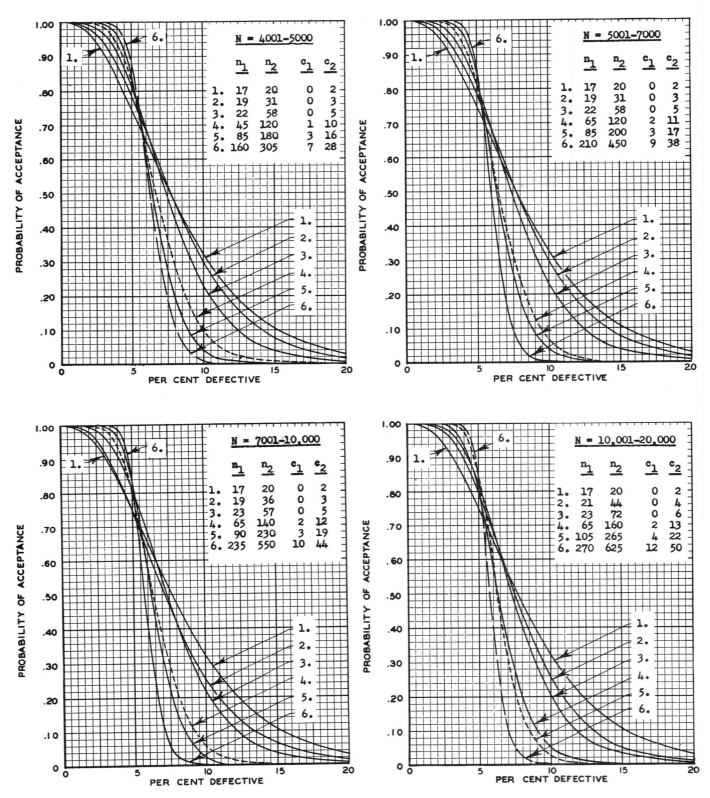

N = 4001-5000

	n_1	n_2	c_1	c_2
1.	17	20	0	2
2.	19	31	0	3
3.	22	58	0	5
4.	45	120	1	10
5.	85	180	3	16
6.	160	305	7	28

N = 5001-7000

	n_1	n_2	c_1	c_2
1.	17	20	0	2
2.	19	31	0	3
3.	22	58	0	5
4.	65	120	2	11
5.	85	200	3	17
6.	210	450	9	38

N = 7001-10,000

	n_1	n_2	c_1	c_2
1.	17	20	0	2
2.	19	36	0	3
3.	23	57	0	5
4.	65	140	2	12
5.	90	230	3	19
6.	235	550	10	44

N = 10,001-20,000

	n_1	n_2	c_1	c_2
1.	17	20	0	2
2.	21	44	0	4
3.	23	72	0	6
4.	65	160	2	13
5.	105	265	4	22
6.	270	625	12	50

PROBABILITY OF ACCEPTANCE

PER CENT DEFECTIVE

Operating Characteristic Curves • Double Sampling Plans
Average Outgoing Quality Limit, AOQL = 4.0%

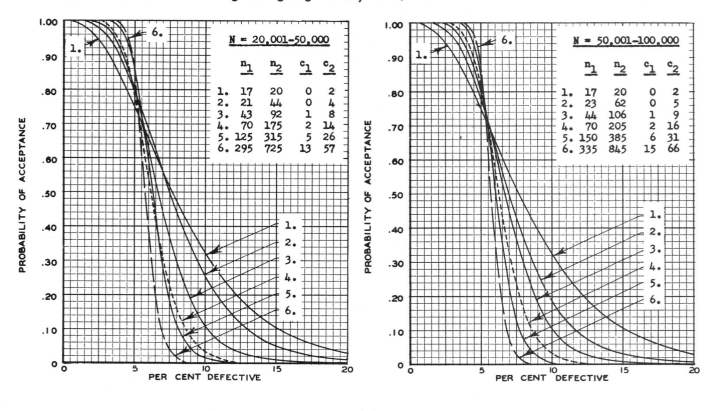

N = 20,001-50,000

	n_1	n_2	c_1	c_2
1.	17	20	0	2
2.	21	44	0	4
3.	43	92	1	8
4.	70	175	2	14
5.	125	315	5	26
6.	295	725	13	57

N = 50,001-100,000

	n_1	n_2	c_1	c_2
1.	17	20	0	2
2.	23	62	0	5
3.	44	106	1	9
4.	70	205	2	16
5.	150	385	6	31
6.	335	845	15	66

Operating Characteristic Curves   •   Double Sampling Plans
Average Outgoing Quality Limit, AOQL = 5.0%

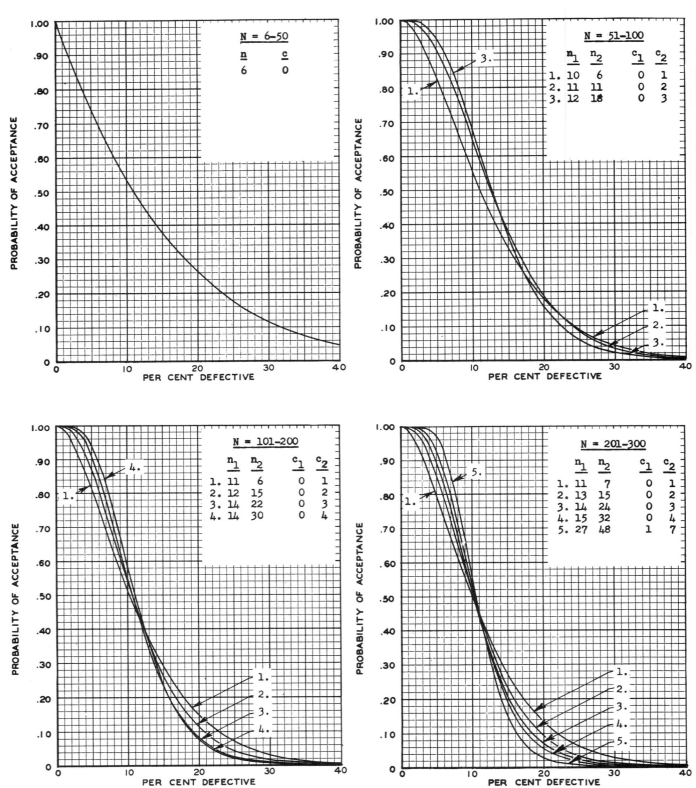

Operating Characteristic Curves • Double Sampling Plans
Average Outgoing Quality Limit, AOQL = 5.0%

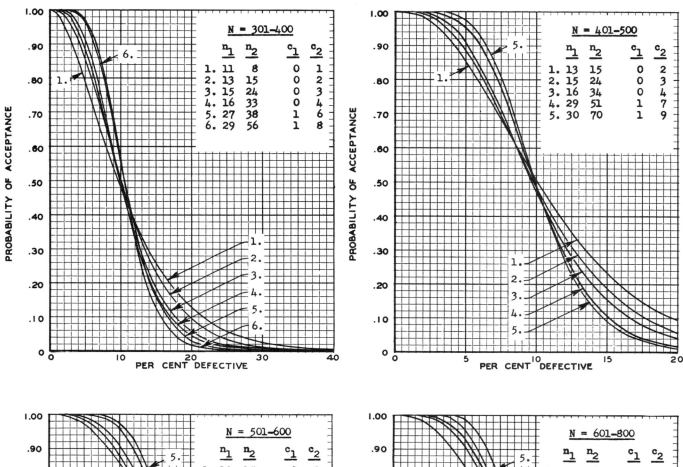

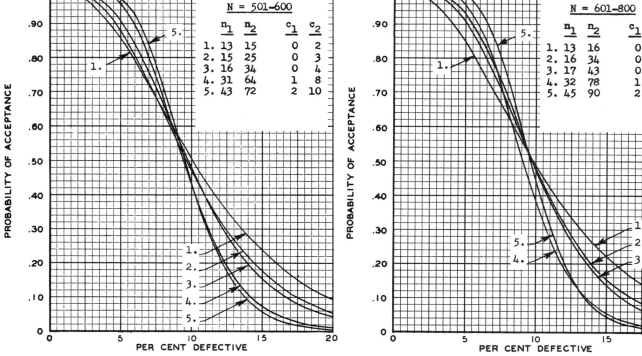

DOUBLE SAMPLING

5.0%

AOQL

Operating Characteristic Curves • Double Sampling Plans
Average Outgoing Quality Limit, AOQL = 5.0%

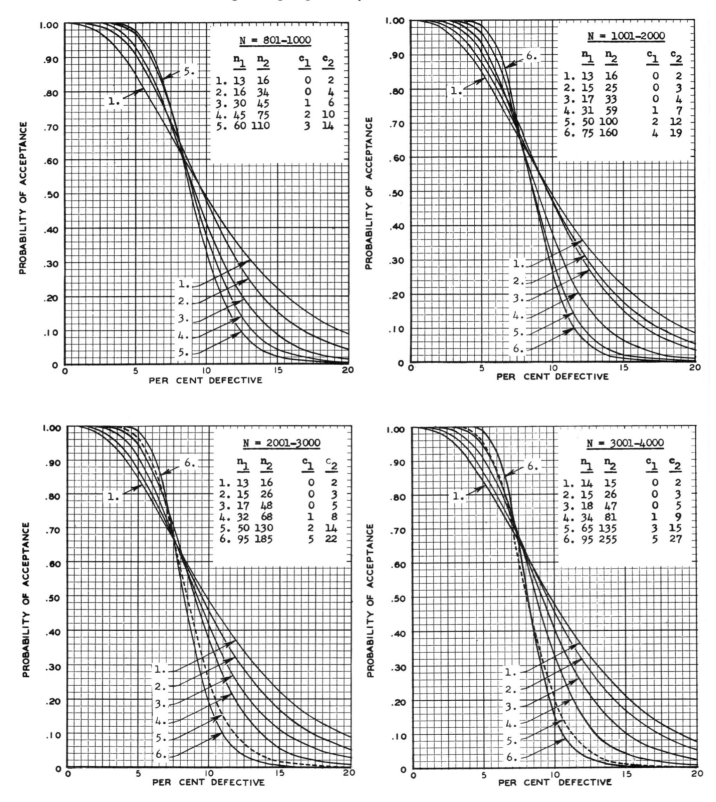

N = 801–1000

	n_1	n_2	c_1	c_2
1.	13	16	0	2
2.	16	34	0	4
3.	30	45	1	6
4.	45	75	2	10
5.	60	110	3	14

N = 1001–2000

	n_1	n_2	c_1	c_2
1.	13	16	0	2
2.	15	25	0	3
3.	17	33	0	4
4.	31	59	1	7
5.	50	100	2	12
6.	75	160	4	19

N = 2001–3000

	n_1	n_2	c_1	c_2
1.	13	16	0	2
2.	15	26	0	3
3.	17	48	0	5
4.	32	68	1	8
5.	50	130	2	14
6.	95	185	5	22

N = 3001–4000

	n_1	n_2	c_1	c_2
1.	14	15	0	2
2.	15	26	0	3
3.	18	47	0	5
4.	34	81	1	9
5.	65	135	3	15
6.	95	255	5	27

PROBABILITY OF ACCEPTANCE

PER CENT DEFECTIVE

Operating Characteristic Curves • Double Sampling Plans
Average Outgoing Quality Limit, AOQL = 5.0%

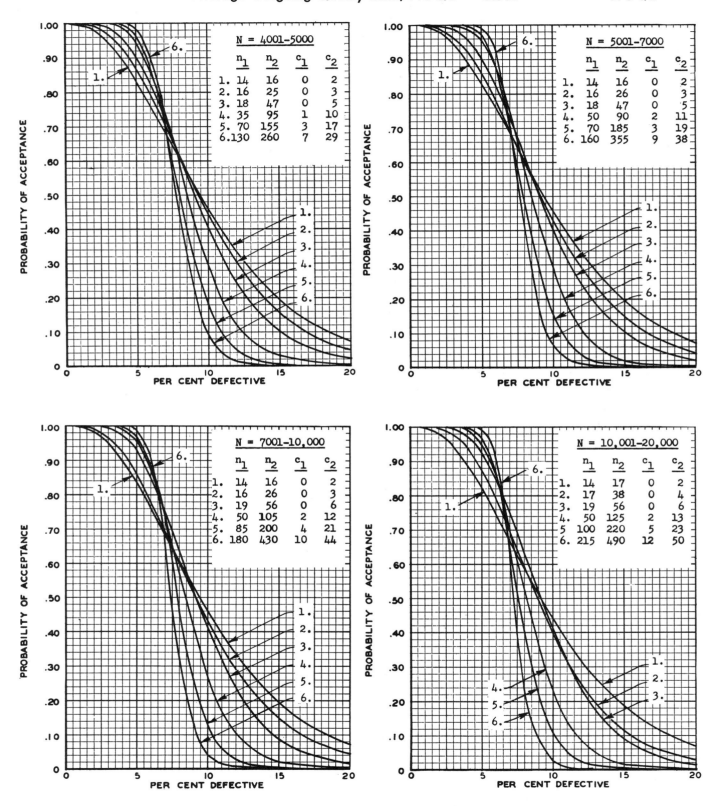

N = 4001-5000

	n_1	n_2	c_1	c_2
1.	14	16	0	2
2.	16	25	0	3
3.	18	47	0	5
4.	35	95	1	10
5.	70	155	3	17
6.	130	260	7	29

N = 5001-7000

	n_1	n_2	c_1	c_2
1.	14	16	0	2
2.	16	26	0	3
3.	18	47	0	5
4.	50	90	2	11
5.	70	185	3	19
6.	160	355	9	38

N = 7001-10,000

	n_1	n_2	c_1	c_2
1.	14	16	0	2
2.	16	26	0	3
3.	19	56	0	6
4.	50	105	2	12
5.	85	200	4	21
6.	180	430	10	44

N = 10,001-20,000

	n_1	n_2	c_1	c_2
1.	14	17	0	2
2.	17	38	0	4
3.	19	56	0	6
4.	50	125	2	13
5.	100	220	5	23
6.	215	490	12	50

PROBABILITY OF ACCEPTANCE

PER CENT DEFECTIVE

DOUBLE
SAMPLING

5.0%

AOQL

Operating Characteristic Curves • Double Sampling Plans

Average Outgoing Quality Limit, AOQL = 5.0%

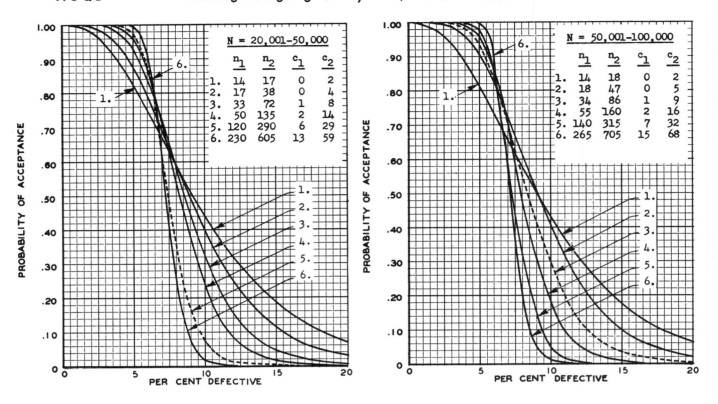

N = 20,001–50,000

	n_1	n_2	c_1	c_2
1.	14	17	0	2
2.	17	38	0	4
3.	33	72	1	8
4.	50	135	2	14
5.	120	290	6	29
6.	230	605	13	59

N = 50,001–100,000

	n_1	n_2	c_1	c_2
1.	14	18	0	2
2.	18	47	0	5
3.	34	86	1	9
4.	55	160	2	16
5.	140	315	7	32
6.	265	705	15	68

Operating Characteristic Curves • Double Sampling Plans
Average Outgoing Quality Limit, AOQL = 7.0%

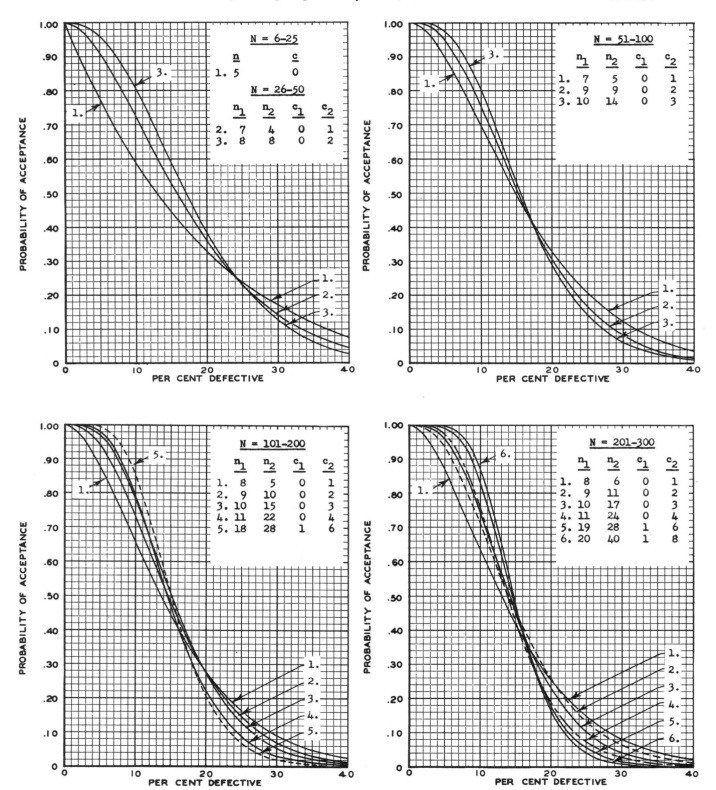

DOUBLE SAMPLING

7.0%

AOQL

Operating Characteristic Curves • Double Sampling Plans
Average Outgoing Quality Limit, AOQL = 7.0%

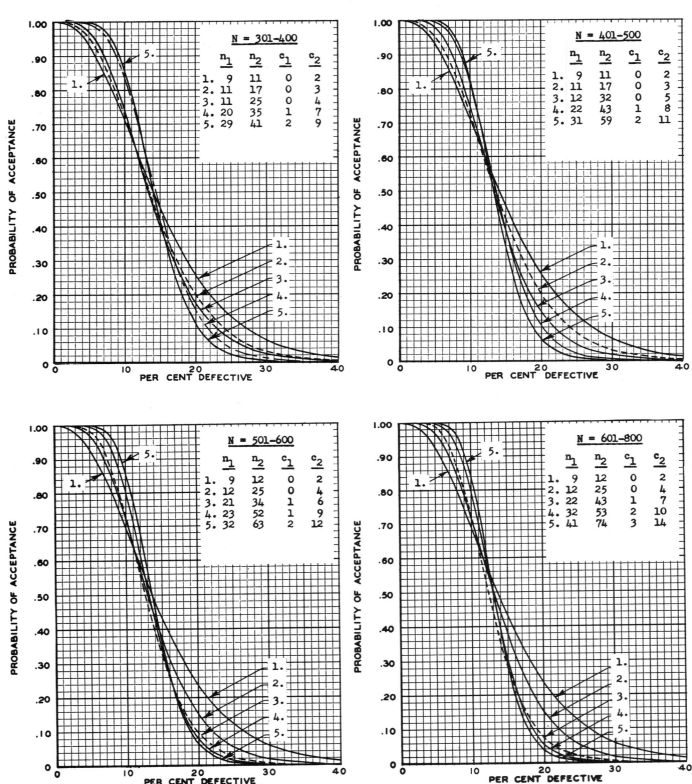

N = 301-400

	n_1	n_2	c_1	c_2
1.	9	11	0	2
2.	11	17	0	3
3.	11	25	0	4
4.	20	35	1	7
5.	29	41	2	9

N = 401-500

	n_1	n_2	c_1	c_2
1.	9	11	0	2
2.	11	17	0	3
3.	12	32	0	5
4.	22	43	1	8
5.	31	59	2	11

N = 501-600

	n_1	n_2	c_1	c_2
1.	9	12	0	2
2.	12	25	0	4
3.	21	34	1	6
4.	23	52	1	9
5.	32	63	2	12

N = 601-800

	n_1	n_2	c_1	c_2
1.	9	12	0	2
2.	12	25	0	4
3.	22	43	1	7
4.	32	53	2	10
5.	41	74	3	14

PROBABILITY OF ACCEPTANCE

PER CENT DEFECTIVE

Operating Characteristic Curves • Double Sampling Plans
Average Outgoing Quality Limit, AOQL = 7.0%

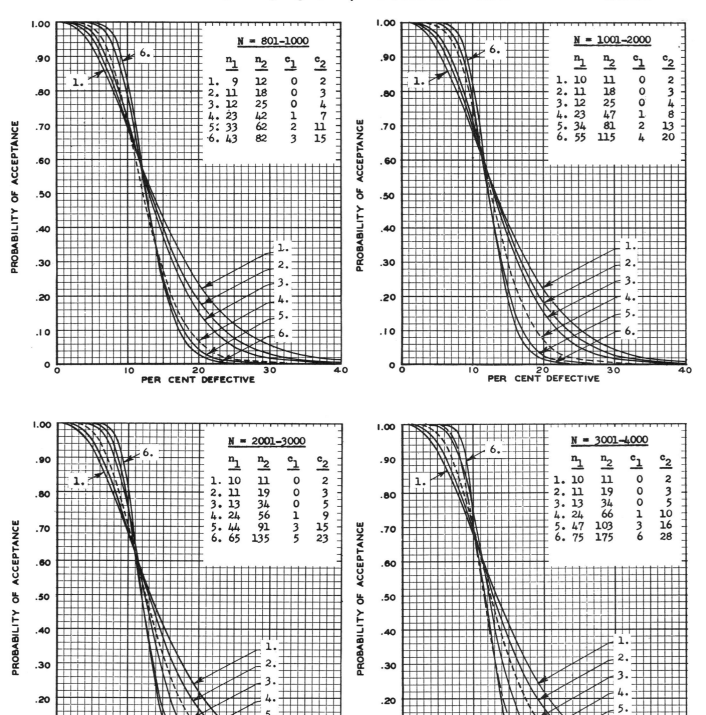

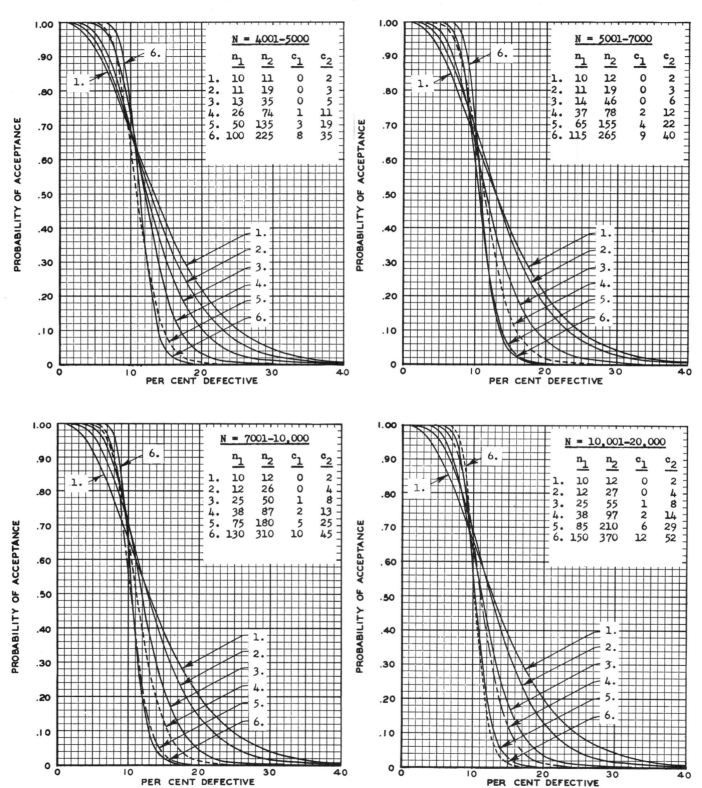

N = 4001-5000

	n_1	n_2	c_1	c_2
1.	10	11	0	2
2.	11	19	0	3
3.	13	35	0	5
4.	26	74	1	11
5.	50	135	3	19
6.	100	225	8	35

N = 5001-7000

	n_1	n_2	c_1	c_2
1.	10	12	0	2
2.	11	19	0	3
3.	14	46	0	6
4.	37	78	2	12
5.	65	155	4	22
6.	115	265	9	40

N = 7001-10,000

	n_1	n_2	c_1	c_2
1.	10	12	0	2
2.	12	26	0	4
3.	25	50	1	8
4.	38	87	2	13
5.	75	180	5	25
6.	130	310	10	45

N = 10,001-20,000

	n_1	n_2	c_1	c_2
1.	10	12	0	2
2.	12	27	0	4
3.	25	55	1	8
4.	38	97	2	14
5.	85	210	6	29
6.	150	370	12	52

PROBABILITY OF ACCEPTANCE

PER CENT DEFECTIVE

Operating Characteristic Curves • Double Sampling Plans
Average Outgoing Quality Limit, AOQL = 7.0%

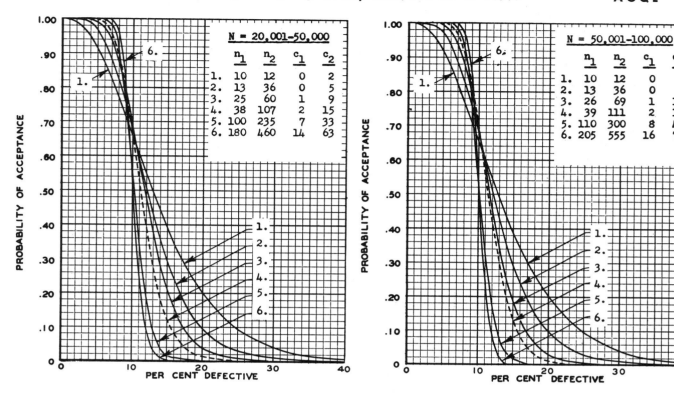

N = 20,001-50,000

	n_1	n_2	c_1	c_2
1.	10	12	0	2
2.	13	36	0	5
3.	25	60	1	9
4.	38	107	2	15
5.	100	235	7	33
6.	180	460	14	63

N = 50,001-100,000

	n_1	n_2	c_1	c_2
1.	10	12	0	2
2.	13	36	0	5
3.	26	69	1	10
4.	39	111	2	16
5.	110	300	8	40
6.	205	555	16	75

DOUBLE
SAMPLING

10.0%

AOQL

— 166 —

Appendix 2

Operating Characteristic Curves • Double Sampling Plans
Average Outgoing Quality Limit, AOQL = 10.0%

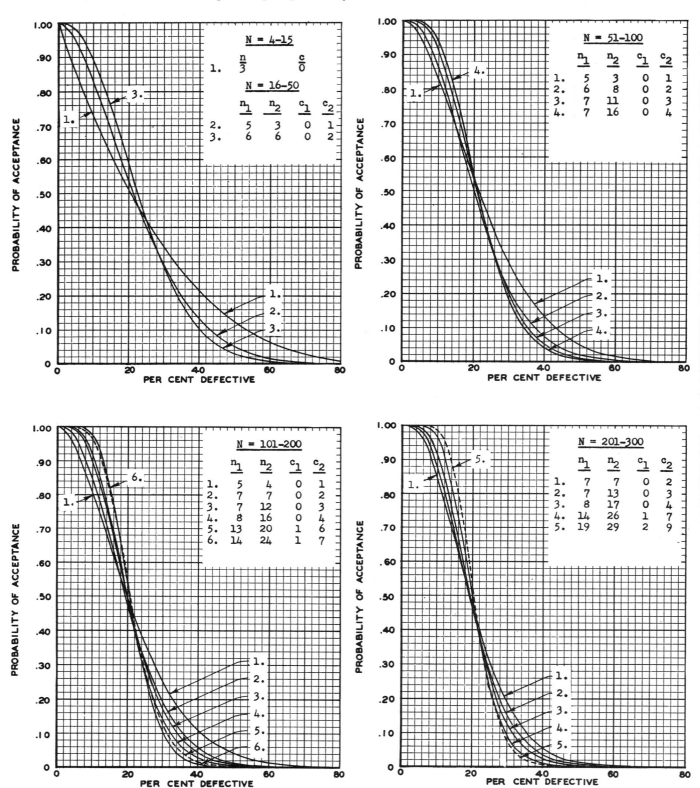

Operating Characteristic Curves • Double Sampling Plans
Average Outgoing Quality Limit, AOQL = 10.0%

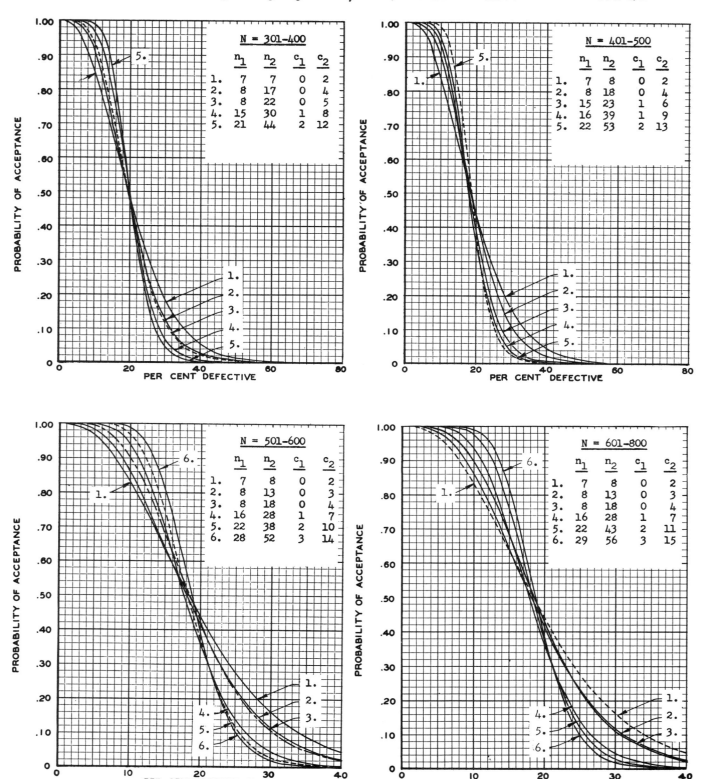

N = 301-400

	n_1	n_2	c_1	c_2
1.	7	7	0	2
2.	8	17	0	4
3.	8	22	0	5
4.	15	30	1	8
5.	21	44	2	12

N = 401-500

	n_1	n_2	c_1	c_2
1.	7	8	0	2
2.	8	18	0	4
3.	15	23	1	6
4.	16	39	1	9
5.	22	53	2	13

N = 501-600

	n_1	n_2	c_1	c_2
1.	7	8	0	2
2.	8	13	0	3
3.	8	18	0	4
4.	16	28	1	7
5.	22	38	2	10
6.	28	52	3	14

N = 601-800

	n_1	n_2	c_1	c_2
1.	7	8	0	2
2.	8	13	0	3
3.	8	18	0	4
4.	16	28	1	7
5.	22	43	2	11
6.	29	56	3	15

PROBABILITY OF ACCEPTANCE

PER CENT DEFECTIVE

DOUBLE
SAMPLING
10.0%
AOQL

Operating Characteristic Curves   •   Double Sampling Plans
Average Outgoing Quality Limit, AOQL = 10.0%

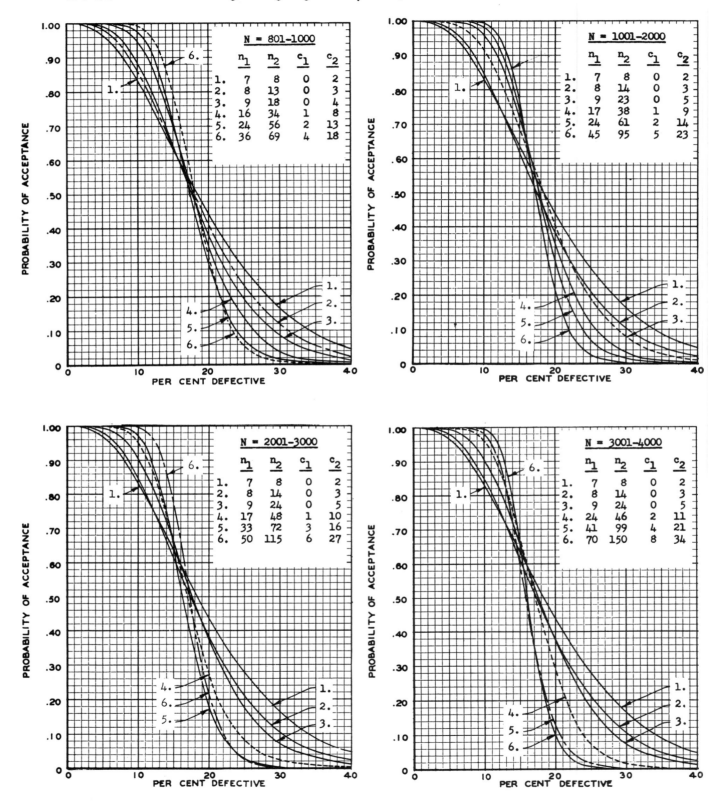

Operating Characteristic Curves • Double Sampling Plans

Average Outgoing Quality Limit, AOQL = 10.0%

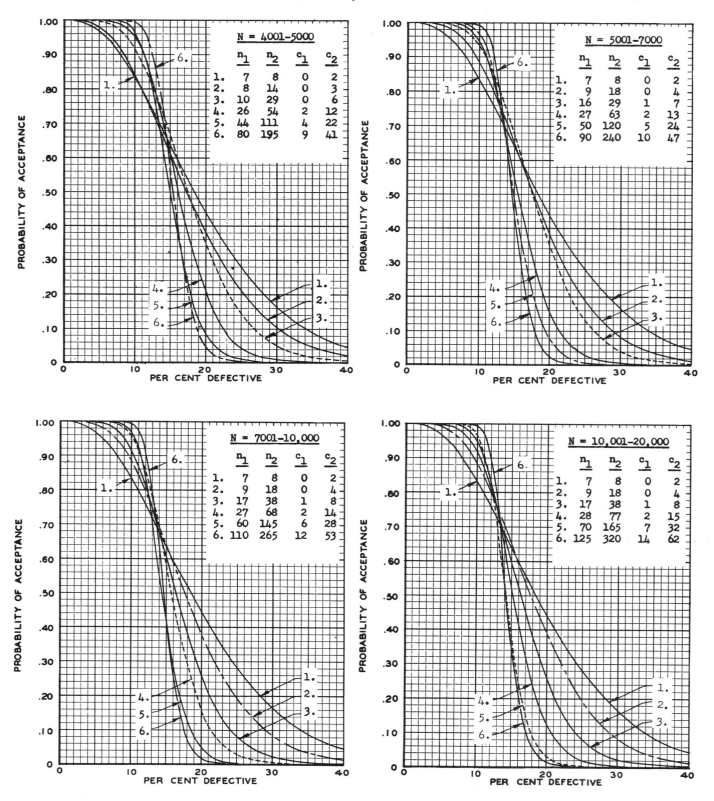

DOUBLE
SAMPLING

10.0%

AOQL

Operating Characteristic Curves • Double Sampling Plans
Average Outgoing Quality Limit, AOQL = 10.0%

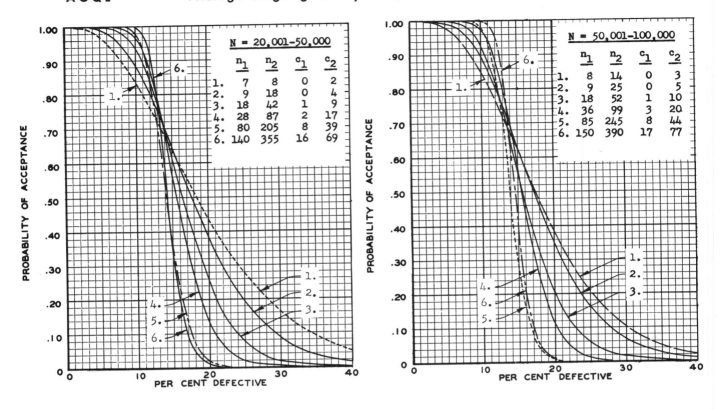

Appendix 3

OC Curves for Single Sampling Plans
with c = 0, 1, 2, 3 and n ≤ 500

n = 2 − 34
c = 0

Operating Characteristic Curves • Single Sampling Plans

Acceptance Number, c = 0

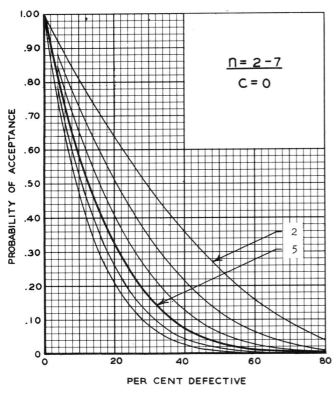

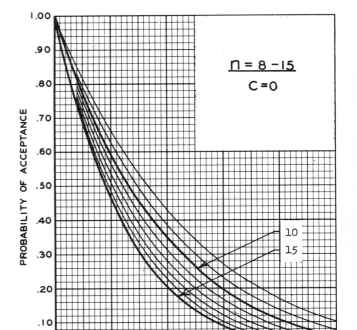

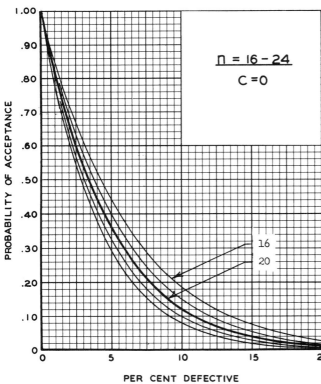

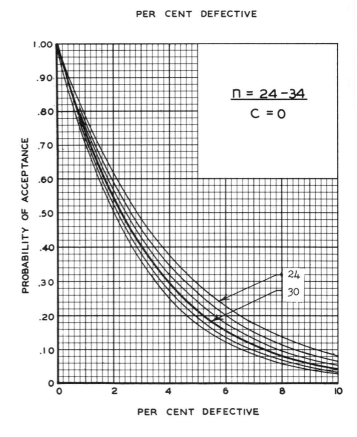

Operating Characteristic Curves • Single Sampling Plans
Acceptance Number, c = 0

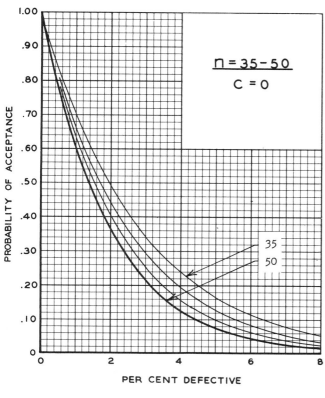

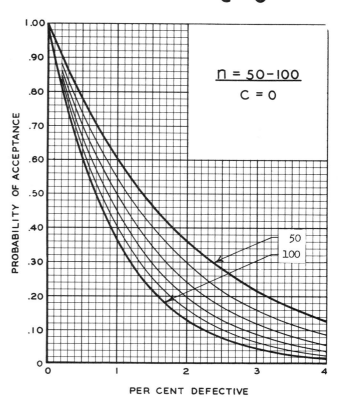

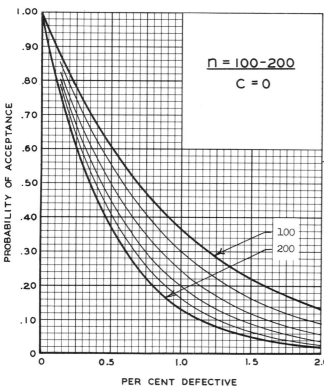

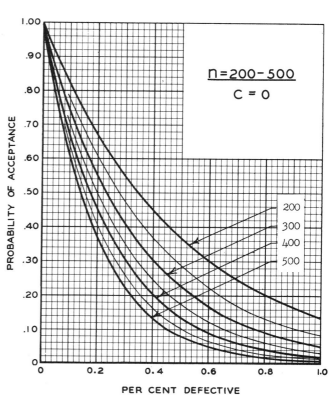

n = 3 − 50
c = 1

Operating Characteristic Curves • Single Sampling Plans
Acceptance Number, c = 1

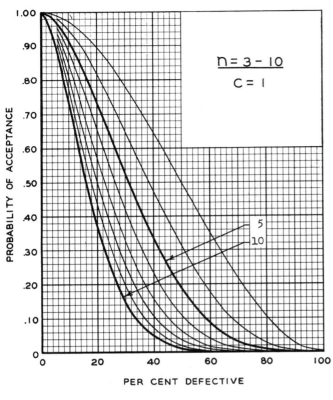

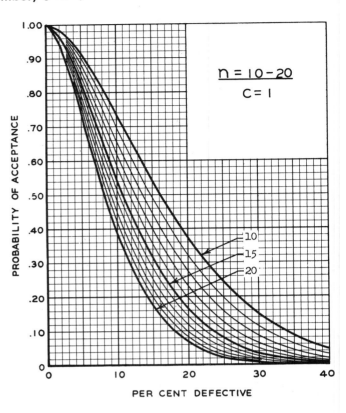

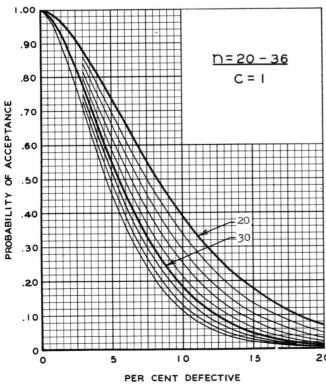

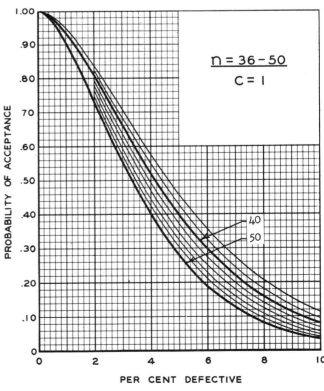

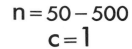

Operating Characteristic Curves • Single Sampling Plans
Acceptance Number, c = 1

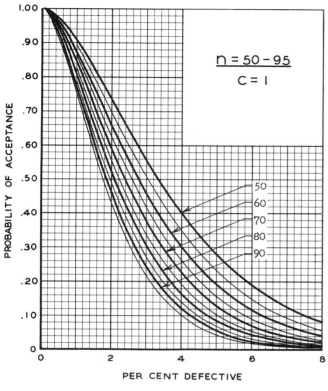

n = 50 – 95
C = 1

PROBABILITY OF ACCEPTANCE

50
60
70
80
90

PER CENT DEFECTIVE

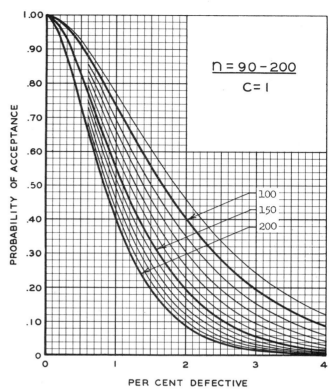

n = 90 – 200
C = 1

PROBABILITY OF ACCEPTANCE

100
150
200

PER CENT DEFECTIVE

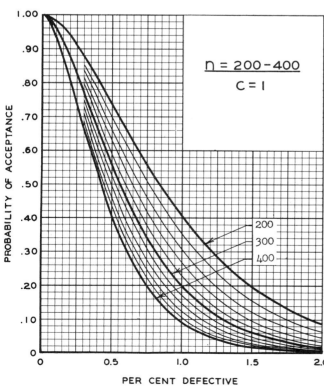

n = 200 – 400
C = 1

PROBABILITY OF ACCEPTANCE

200
300
400

PER CENT DEFECTIVE

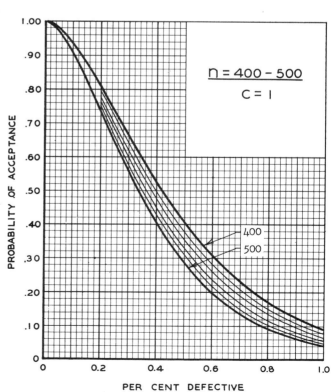

n = 400 – 500
C = 1

PROBABILITY OF ACCEPTANCE

400
500

PER CENT DEFECTIVE

SINGLE
SAMPLING

n = 5 – 60
c = 2

Operating Characteristic Curves • Single Sampling Plans

Acceptance Number, c = 2

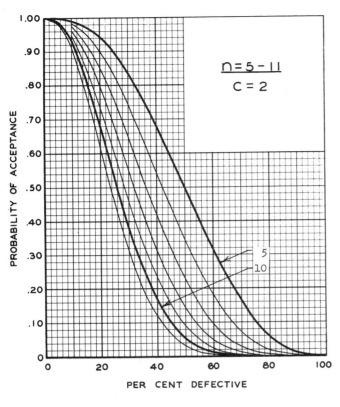

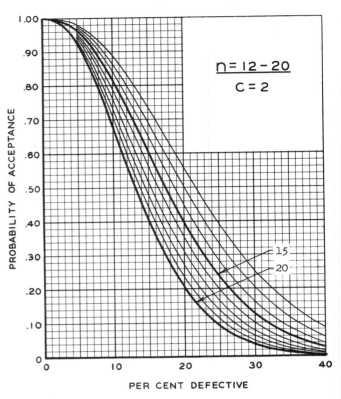

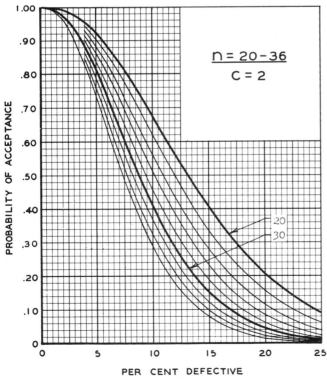

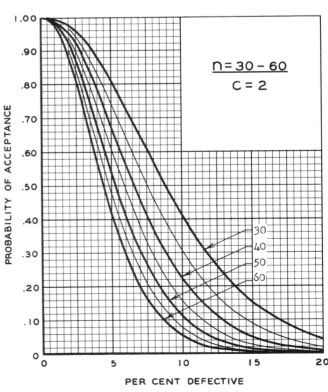

Operating Characteristic Curves • Single Sampling Plans

Acceptance Number, c = 2

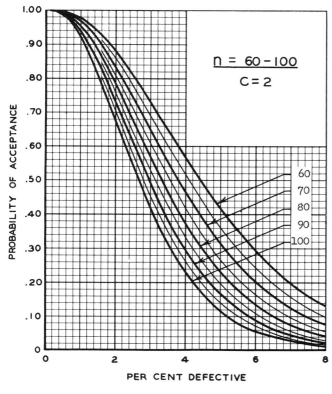

n = 60 − 100
C = 2

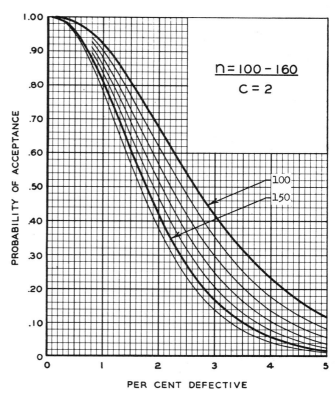

n = 100 − 160
C = 2

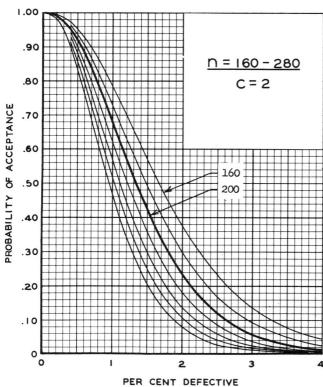

n = 160 − 280
C = 2

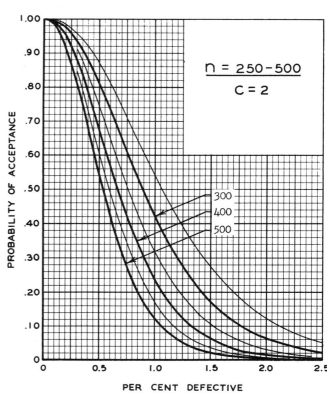

n = 250 − 500
C = 2

n = 8−75
c = 3

Operating Characteristic Curves • Single Sampling Plans
Acceptance Number, c = 3

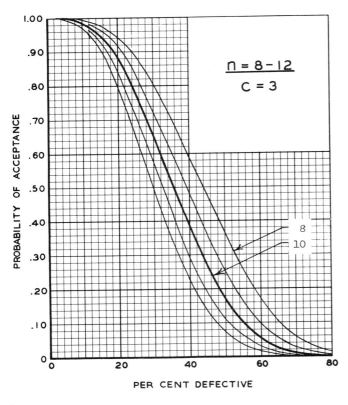

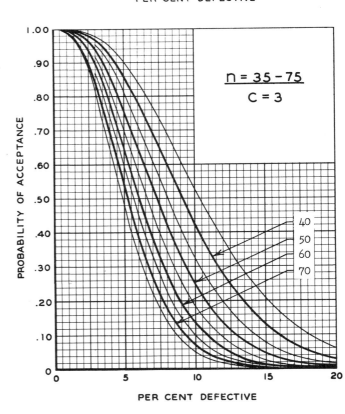

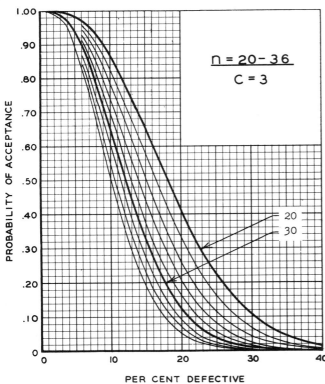

Operating Characteristic Curves   •   Single Sampling Plans   **n = 70 − 500**

Acceptance Number, c = 3   **c = 3**

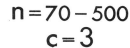

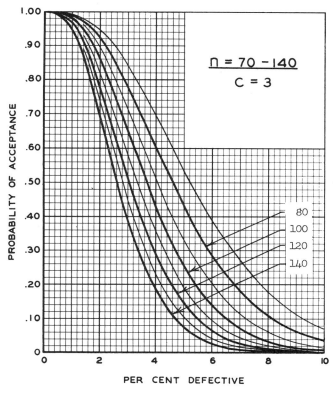

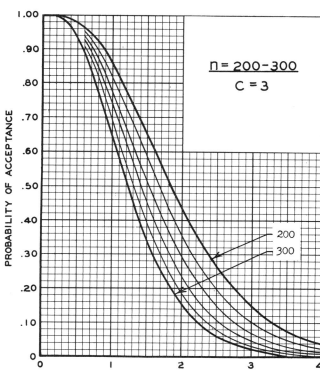

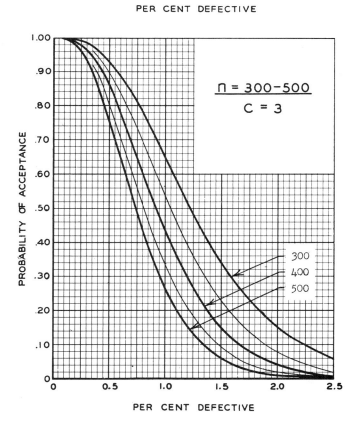

Appendix 4

Single Sampling Tables for Stated Values of
Lot Tolerance Per Cent Defective (LTPD)
with Consumer's Risk of 0.10

0.5% LTPD

Single Sampling Table for Lot Tolerance Per Cent Defective (LTPD) = 0.5%

Lot Size	Process Average 0 to 0.005%			Process Average 0.006 to 0.050%			Process Average 0.051 to 0.100%			Process Average 0.101 to 0.150%			Process Average 0.151 to 0.200%			Process Average 0.201 to 0.250%		
	n	c	AOQL %	n	c	AOQL %	n	c	AOQL %	n	c	AOQL %	n	c	AOQL %	n	c	AOQL %
1–180	All	0	0	All	0	0	All	0	0	All	0	0	All	0	0	All	0	0
181–210	180	0	0.02	180	0	0.02	180	0	0.02	180	0	0.02	180	0	0.02	180	0	0.02
211–250	210	0	0.03	210	0	0.03	210	0	0.03	210	0	0.03	210	0	0.03	210	0	0.03
251–300	240	0	0.03	240	0	0.03	240	0	0.03	240	0	0.03	240	0	0.03	240	0	0.03
301–400	275	0	0.04	275	0	0.04	275	0	0.04	275	0	0.04	275	0	0.04	275	0	0.04
401–500	300	0	0.05	300	0	0.05	300	0	0.05	300	0	0.05	300	0	0.05	300	0	0.05
501–600	320	0	0.05	320	0	0.05	320	0	0.05	320	0	0.05	320	0	0.05	320	0	0.05
601–800	350	0	0.06	350	0	0.06	350	0	0.06	350	0	0.06	350	0	0.06	350	0	0.06
801–1000	365	0	0.06	365	0	0.06	365	0	0.06	365	0	0.06	365	0	0.06	365	0	0.06
1001–2000	410	0	0.07	410	0	0.07	410	0	0.07	670	1	0.08	670	1	0.08	670	1	0.08
2001–3000	430	0	0.07	430	0	0.07	705	1	0.09	705	1	0.09	955	2	0.10	955	2	0.10
3001–4000	440	0	0.07	440	0	0.07	730	1	0.09	985	2	0.10	1230	3	0.11	1230	3	0.11
4001–5000	445	0	0.08	740	1	0.10	1000	2	0.11	1000	2	0.11	1250	3	0.12	1480	4	0.12
5001–7000	450	0	0.08	750	1	0.10	1020	2	0.12	1280	3	0.12	1510	4	0.13	1760	5	0.14
7001–10,000	455	0	0.08	760	1	0.10	1040	2	0.12	1530	4	0.14	1790	5	0.14	2240	7	0.16
10,001–20,000	460	0	0.08	775	1	0.10	1330	3	0.14	1820	5	0.16	2300	7	0.17	2780	9	0.18
20,001–50,000	775	1	0.11	1050	2	0.13	1600	4	0.15	2080	5	0.18	3060	10	0.20	4200	15	0.22
50,001–100,000	780	1	0.11	1060	2	0.13	1840	5	0.17	2590	8	0.19	3780	13	0.22	5140	19	0.24

1.0%

Single Sampling Table for Lot Tolerance Per Cent Defective (LTPD) = 1.0%

Lot Size	Process Average 0 to 0.010%			Process Average 0.011 to 0.10%			Process Average 0.11 to 0.20%			Process Average 0.21 to 0.30%			Process Average 0.31 to 0.40%			Process Average 0.41 to 0.50%		
	n	c	AOQL %	n	c	AOQL %	n	c	AOQL %	n	c	AOQL %	n	c	AOQL %	n	c	AOQL %
1–120	All	0	0	All	0	0	All	0	0	All	0	0	All	0	0	All	0	0
121–150	120	0	0.06	120	0	0.06	120	0	0.06	120	0	0.06	120	0	0.06	120	0	0.06
151–200	140	0	0.08	140	0	0.08	140	0	0.08	140	0	0.08	140	0	0.08	140	0	0.08
201–300	165	0	0.10	165	0	0.10	165	0	0.10	165	0	0.10	165	0	0.10	165	0	0.10
301–400	175	0	0.12	175	0	0.12	175	0	0.12	175	0	0.12	175	0	0.12	175	0	0.12
401–500	180	0	0.13	180	0	0.13	180	0	0.13	180	0	0.13	180	0	0.13	180	0	0.13
501–600	190	0	0.13	190	0	0.13	190	0	0.13	190	0	0.13	190	0	0.13	305	1	0.14
601–800	200	0	0.14	200	0	0.14	200	0	0.14	330	1	0.15	330	1	0.15	330	1	0.15
801–1000	205	0	0.14	205	0	0.14	205	0	0.14	335	1	0.17	335	1	0.17	335	1	0.17
1001–2000	220	0	0.15	220	0	0.15	360	1	0.19	490	2	0.21	490	2	0.21	610	3	0.22
2001–3000	220	0	0.15	375	1	0.20	505	2	0.23	630	3	0.24	745	4	0.26	870	5	0.26
3001–4000	225	0	0.15	380	1	0.20	510	2	0.24	645	3	0.25	880	5	0.28	1000	6	0.29
4001–5000	225	0	0.16	380	1	0.20	520	2	0.24	770	4	0.28	895	5	0.29	1120	7	0.31
5001–7000	230	0	0.15	385	1	0.21	655	3	0.27	780	4	0.29	1020	6	0.32	1260	8	0.34
7001–10,000	230	0	0.16	520	2	0.25	660	3	0.28	910	5	0.32	1150	7	0.34	1500	10	0.37
10,001–20,000	390	1	0.21	525	2	0.26	785	4	0.31	1040	6	0.35	1400	9	0.39	1980	14	0.43
20,001–50,000	390	1	0.21	530	2	0.26	920	5	0.34	1300	8	0.39	1890	13	0.44	2570	19	0.48
50,001–100,000	390	1	0.21	670	3	0.29	1040	6	0.36	1420	9	0.41	2120	15	0.47	3150	23	0.50

n = sample size; c = acceptance number
"All" indicates that each piece in the lot is to be inspected
AOQL = Average Outgoing Quality Limit

Single Sampling Table for
Lot Tolerance Per Cent Defective (LTPD) = 2.0%

Lot Size	Process Average 0 to 0.02%			Process Average 0.03 to 0.20%			Process Average 0.21 to 0.40%			Process Average 0.41 to 0.60%			Process Average 0.61 to 0.80%			Process Average 0.81 to 1.00%		
	n	c	AOQL %	n	c	AOQL %	n	c	AOQL %	n	c	AOQL %	n	c	AOQL %	n	c	AOQL %
1–75	All	0	0	All	0	0	All	0	0	All	0	0	All	0	0	All	0	0
76–100	70	0	0.16	70	0	0.16	70	0	0.16	70	0	0.16	70	0	0.16	70	0	0.16
101–200	85	0	0.25	85	0	0.25	85	0	0.25	85	0	0.25	85	0	0.25	85	0	0.25
201–300	95	0	0.26	95	0	0.26	95	0	0.26	95	0	0.26	95	0	0.26	95	0	0.26
301–400	100	0	0.28	100	0	0.28	100	0	0.28	160	1	0.32	160	1	0.32	160	1	0.32
401–500	105	0	0.28	105	0	0.28	105	0	0.28	165	1	0.34	165	1	0.34	165	1	0.34
501–600	105	0	0.29	105	0	0.29	175	1	0.34	175	1	0.34	175	1	0.34	235	2	0.36
601–800	110	0	0.29	110	0	0.29	180	1	0.36	240	2	0.40	240	2	0.40	300	3	0.41
801–1000	115	0	0.28	115	0	0.28	185	1	0.37	245	2	0.42	305	3	0.44	305	3	0.44
1001–2000	115	0	0.30	190	1	0.40	255	2	0.47	325	3	0.50	380	4	0.54	440	5	0.56
2001–3000	115	0	0.31	190	1	0.41	260	2	0.48	385	4	0.58	450	5	0.60	565	7	0.64
3001–4000	115	0	0.31	195	1	0.41	330	3	0.54	450	5	0.63	510	6	0.65	690	9	0.70
4001–5000	195	1	0.41	260	2	0.50	335	3	0.54	455	5	0.63	575	7	0.69	750	10	0.74
5001–7000	195	1	0.42	265	2	0.50	335	3	0.55	515	6	0.69	640	8	0.73	870	12	0.80
7001–10,000	195	1	0.42	265	2	0.50	395	4	0.62	520	6	0.69	760	10	0.79	1050	15	0.86
10,001–20,000	200	1	0.42	265	2	0.51	460	5	0.67	650	8	0.77	885	12	0.86	1230	18	0.94
20,001–50,000	200	1	0.42	335	3	0.58	520	6	0.73	710	9	0.81	1060	15	0.93	1520	23	1.0
50,001–100,000	200	1	0.42	335	3	0.58	585	7	0.76	770	10	0.84	1180	17	0.97	1690	26	1.1

Single Sampling Table for
Lot Tolerance Per Cent Defective (LTPD) = 3.0%

3.0%

Lot Size	Process Average 0 to 0.03%			Process Average 0.04 to 0.30%			Process Average 0.31 to 0.60%			Process Average 0.61 to 0.90%			Process Average 0.91 to 1.20%			Process Average 1.21 to 1.50%		
	n	c	AOQL %	n	c	AOQL %	n	c	AOQL %	n	c	AOQL %	n	c	AOQL %	n	c	AOQL %
1–40	All	0	0	All	0	0	All	0	0	All	0	0	All	0	0	All	0	0
41–55	40	0	0.18	40	0	0.18	40	0	0.18	40	0	0.18	40	0	0.18	40	0	0.18
56–100	55	0	0.30	55	0	0.30	55	0	0.30	55	0	0.30	55	0	0.30	55	0	0.30
101–200	65	0	0.38	65	0	0.38	65	0	0.38	65	0	0.38	65	0	0.38	65	0	0.38
201–300	70	0	0.40	70	0	0.40	70	0	0.40	110	1	0.48	110	1	0.48	110	1	0.48
301–400	70	0	0.43	70	0	0.43	115	1	0.52	115	1	0.52	115	1	0.52	155	2	0.54
401–500	70	0	0.45	70	0	0.45	120	1	0.53	120	1	0.53	160	2	0.58	160	2	0.58
501–600	75	0	0.43	75	0	0.43	120	1	0.56	160	2	0.63	160	2	0.63	200	3	0.65
601–800	75	0	0.44	125	1	0.57	125	1	0.57	165	2	0.66	205	3	0.71	240	4	0.74
801–1000	75	0	0.45	125	1	0.59	170	2	0.67	210	3	0.73	250	4	0.76	290	5	0.78
1001–2000	75	0	0.47	130	1	0.60	175	2	0.72	260	4	0.85	300	5	0.90	380	7	0.95
2001–3000	75	0	0.48	130	1	0.62	220	3	0.82	300	5	0.95	385	7	1.0	460	9	1.1
3001–4000	130	1	0.63	175	2	0.75	220	3	0.84	305	5	0.96	425	8	1.1	540	11	1.2
4001–5000	130	1	0.63	175	2	0.76	260	4	0.91	345	6	1.0	465	9	1.1	620	13	1.2
5001–7000	130	1	0.63	175	2	0.76	265	4	0.92	390	7	1.1	505	10	1.2	700	15	1.3
7001–10,000	130	1	0.64	175	2	0.77	265	4	0.93	390	7	1.1	550	11	1.2	775	17	1.4
10,001–20,000	130	1	0.64	175	2	0.78	305	5	1.0	430	8	1.2	630	13	1.3	900	20	1.5
20,001–50,000	130	1	0.65	225	3	0.86	350	6	1.1	520	10	1.2	750	16	1.4	1090	25	1.6
50,001–100,000	130	1	0.65	265	4	0.96	390	7	1.1	590	12	1.3	830	18	1.5	1215	28	1.6

n = sample size; c = acceptance number
"All" indicates that each piece in the lot is to be inspected
AOQL = Average Outgoing Quality Limit

SINGLE
SAMPLING

4.0%
LTPD

— 184 —

Appendix 4

Single Sampling Table for
Lot Tolerance Per Cent Defective (LTPD) = 4.0%

Lot Size	Process Average 0 to 0.04%			Process Average 0.05 to 0.40%			Process Average 0.41 to 0.80%			Process Average 0.81 to 1.20%			Process Average 1.21 to 1.60%			Process Average 1.61 to 2.00%		
	n	c	AOQL %	n	c	AOQL %	n	c	AOQL %	n	c	AOQL %	n	c	AOQL %	n	c	AOQL %
1–35	All	0	0	All	0	0	All	0	0	All	0	0	All	0	0	All	0	0
36–50	34	0	0.35	34	0	0.35	34	0	0.35	34	0	0.35	34	0	0.35	34	0	0.35
51–100	44	0	0.47	44	0	0.47	44	0	0.47	44	0	0.47	44	0	0.47	44	0	0.47
101–200	50	0	0.55	50	0	0.55	50	0	0.55	50	0	0.55	50	0	0.55	50	0	0.55
201–300	55	0	0.57	55	0	0.57	85	1	0.71	85	1	0.71	85	1	0.71	85	1	0.71
301–400	55	0	0.58	55	0	0.58	90	1	0.72	120	2	0.80	120	2	0.80	145	3	0.86
401–500	55	0	0.60	55	0	0.60	90	1	0.77	120	2	0.87	150	3	0.91	150	3	0.91
501–600	55	0	0.61	95	1	0.76	125	2	0.87	125	2	0.87	155	3	0.93	185	4	0.95
601–800	55	0	0.62	95	1	0.78	125	2	0.93	160	3	0.97	190	4	1.0	220	5	1.0
801–1000	55	0	0.63	95	1	0.80	130	2	0.92	165	3	0.98	220	5	1.1	255	6	1.1
1001–2000	55	0	0.65	95	1	0.84	165	3	1.1	195	4	1.2	255	6	1.3	315	8	1.4
2001–3000	95	1	0.86	130	2	1.0	165	3	1.1	230	5	1.3	320	8	1.4	405	11	1.6
3001–4000	95	1	0.86	130	2	1.0	195	4	1.2	260	6	1.4	350	9	1.5	465	13	1.6
4001–5000	95	1	0.87	130	2	1.0	195	4	1.2	290	7	1.4	380	10	1.6	520	15	1.7
5001–7000	95	1	0.87	130	2	1.0	200	4	1.2	290	7	1.5	410	11	1.7	575	17	1.9
7001–10,000	95	1	0.88	130	2	1.1	230	5	1.4	325	8	1.5	440	12	1.7	645	19	1.9
10,001–20,000	95	1	0.88	165	3	1.2	265	6	1.4	355	9	1.6	500	14	1.8	730	22	2.0
20,001–50,000	95	1	0.88	165	3	1.2	295	7	1.5	380	10	1.7	590	17	2.0	870	26	2.1
50,001–100,000	95	1	0.88	200	4	1.3	325	8	1.6	410	11	1.8	620	18	2.0	925	29	2.2

5.0%

Single Sampling Table for
Lot Tolerance Per Cent Defective (LTPD) = 5.0%

Lot Size	Process Average 0 to 0.05%			Process Average 0.06 to 0.50%			Process Average 0.51 to 1.00%			Process Average 1.01 to 1.50%			Process Average 1.51 to 2.00%			Process Average 2.01 to 2.50%		
	n	c	AOQL %	n	c	AOQL %	n	c	AOQL %	n	c	AOQL %	n	c	AOQL %	n	c	AOQL %
1–30	All	0	0	All	0	0	All	0	0	All	0	0	All	0	0	All	0	0
31–50	30	0	0.49	30	0	0.49	30	0	0.49	30	0	0.49	30	0	0.49	30	0	0.49
51–100	37	0	0.63	37	0	0.63	37	0	0.63	37	0	0.63	37	0	0.63	37	0	0.63
101–200	40	0	0.74	40	0	0.74	40	0	0.74	40	0	0.74	40	0	0.74	40	0	0.74
201–300	43	0	0.74	43	0	0.74	70	1	0.92	70	1	0.92	95	2	0.99	95	2	0.99
301–400	44	0	0.74	44	0	0.74	70	1	0.99	100	2	1.0	120	3	1.1	145	4	1.1
401–500	45	0	0.75	75	1	0.95	100	2	1.1	100	2	1.1	125	3	1.2	150	4	1.2
501–600	45	0	0.76	75	1	0.98	100	2	1.1	125	3	1.2	150	4	1.3	175	5	1.3
601–800	45	0	0.77	75	1	1.0	100	2	1.2	130	3	1.2	175	5	1.4	200	6	1.4
801–1000	45	0	0.78	75	1	1.0	105	2	1.2	155	4	1.4	180	5	1.4	225	7	1.5
1001–2000	45	0	0.80	75	1	1.0	130	3	1.4	180	5	1.6	230	7	1.7	280	9	1.8
2001–3000	75	1	1.1	105	2	1.3	135	3	1.4	210	6	1.7	280	9	1.9	370	13	2.1
3001–4000	75	1	1.1	105	2	1.3	160	4	1.5	210	6	1.7	305	10	2.0	420	15	2.2
4001–5000	75	1	1.1	105	2	1.3	160	4	1.5	235	7	1.8	330	11	2.0	440	16	2.2
5001–7000	75	1	1.1	105	2	1.3	185	5	1.7	260	8	1.9	350	12	2.2	490	18	2.4
7001–10,000	75	1	1.1	105	2	1.3	185	5	1.7	260	8	1.9	380	13	2.2	535	20	2.5
10,001–20,000	75	1	1.1	135	3	1.4	210	6	1.8	285	9	2.0	425	15	2.3	610	23	2.6
20,001–50,000	75	1	1.1	135	3	1.4	235	7	1.9	305	10	2.1	470	17	2.4	700	27	2.7
50,001–100,000	75	1	1.1	160	4	1.6	235	7	1.9	355	12	2.2	515	19	2.5	770	30	2.8

n = sample size; c = acceptance number
"All" indicates that each piece in the lot is to be inspected
AOQL = Average Outgoing Quality Limit

Single Sampling Table for Lot Tolerance Per Cent Defective (LTPD) = 7.0%

Lot Size	Process Average 0 to 0.07%			Process Average 0.08 to 0.70%			Process Average 0.71 to 1.40%			Process Average 1.41 to 2.10%			Process Average 2.11 to 2.80%			Process Average 2.81 to 3.50%		
	n	c	AOQL %	n	c	AOQL %	n	c	AOQL %	n	c	AOQL %	n	c	AOQL %	n	c	AOQL %
1–25	All	0	0	All	0	0	All	0	0	All	0	0	All	0	0	All	0	0
26–50	24	0	0.80	24	0	0.80	24	0	0.80	24	0	0.80	24	0	0.80	24	0	0.80
51–100	28	0	0.95	28	0	0.95	28	0	0.95	28	0	0.95	28	0	0.95	28	0	0.95
101–200	30	0	1.0	30	0	1.0	49	1	1.3	49	1	1.3	49	1	1.3	65	2	1.4
201–300	31	0	1.1	31	0	1.1	50	1	1.4	70	2	1.5	85	3	1.6	85	3	1.6
301–400	32	0	1.1	55	1	1.4	70	2	1.6	90	3	1.7	105	4	1.8	125	5	1.8
401–500	32	0	1.1	55	1	1.4	75	2	1.6	90	3	1.8	110	4	1.9	140	6	2.0
501–600	32	0	1.1	55	1	1.4	75	2	1.7	95	3	1.8	125	5	2.0	145	6	2.1
601–800	32	0	1.1	55	1	1.4	75	2	1.7	110	4	2.0	130	5	2.1	160	7	2.2
801–1000	33	0	1.1	55	1	1.4	95	3	1.9	110	4	2.1	145	6	2.2	180	8	2.4
1001–2000	55	1	1.5	75	2	1.8	95	3	2.0	130	5	2.3	185	8	2.5	230	11	2.8
2001–3000	55	1	1.5	75	2	1.8	115	4	2.1	150	6	2.4	215	10	2.8	300	15	3.0
3001–4000	55	1	1.5	75	2	1.8	115	4	2.2	165	7	2.6	235	11	2.9	330	17	3.2
4001–5000	55	1	1.5	75	2	1.8	130	5	2.4	185	8	2.7	250	12	3.0	350	18	3.3
5001–7000	55	1	1.5	75	2	1.8	130	5	2.4	185	8	2.7	270	13	3.1	385	20	3.4
7001–10,000	55	1	1.5	95	3	2.0	150	6	2.5	200	9	2.9	285	14	3.2	415	22	3.6
10,001–20,000	55	1	1.5	95	3	2.0	150	6	2.5	220	10	2.9	320	16	3.3	470	25	3.7
20,001–50,000	55	1	1.5	115	4	2.2	170	7	2.6	235	11	3.1	355	18	3.5	530	29	3.9
50,001–100,000	55	1	1.5	115	4	2.2	185	8	2.7	270	13	3.1	370	19	3.5	530	29	3.9

Single Sampling Table for Lot Tolerance Per Cent Defective (LTPD) = 10.0%

Lot Size	Process Average 0 to 0.10%			Process Average 0.11 to 1.00%			Process Average 1.01 to 2.00%			Process Average 2.01 to 3.00%			Process Average 3.01 to 4.00%			Process Average 4.01 to 5.00%		
	n	c	AOQL %	n	c	AOQL %	n	c	AOQL %	n	c	AOQL %	n	c	AOQL %	n	c	AOQL %
1–20	All	0	0	All	0	0	All	0	0	All	0	0	All	0	0	All	0	0
21–50	17	0	1.3	17	0	1.3	17	0	1.3	17	0	1.3	17	0	1.3	17	0	1.3
51–100	20	0	1.5	20	0	1.5	20	0	1.5	33	1	1.7	33	1	1.7	33	1	1.7
101–200	22	0	1.5	22	0	1.5	35	1	2.0	48	2	2.2	48	2	2.2	60	3	2.4
201–300	23	0	1.5	38	1	1.9	50	2	2.3	65	3	2.4	75	4	2.6	85	5	2.7
301–400	23	0	1.5	38	1	2.0	50	2	2.4	65	3	2.5	90	5	2.7	100	6	2.9
401–500	23	0	1.5	38	1	2.0	50	2	2.5	75	4	2.8	90	5	2.9	110	7	3.2
501–600	23	0	1.5	38	1	2.1	65	3	2.7	80	4	3.0	100	6	3.2	125	8	3.3
601–800	23	0	1.6	38	1	2.1	65	3	2.8	90	5	3.1	100	6	3.3	140	9	3.4
801–1000	39	1	2.1	50	2	2.6	65	3	2.8	90	5	3.2	115	7	3.4	150	10	3.7
1001–2000	39	1	2.1	50	2	2.6	80	4	3.1	105	6	3.4	140	9	3.9	195	14	4.4
2001–3000	39	1	2.1	50	2	2.6	80	4	3.1	115	7	3.7	165	11	4.1	230	17	4.7
3001–4000	39	1	2.1	50	2	2.6	90	5	3.4	130	8	3.8	190	13	4.4	255	19	4.8
4001–5000	39	1	2.1	50	2	2.6	90	5	3.5	130	8	3.9	200	14	4.5	270	20	4.9
5001–7000	39	1	2.1	65	3	3.0	105	6	3.6	140	9	4.1	200	14	4.6	295	22	5.0
7001–10,000	39	1	2.2	65	3	3.0	105	6	3.6	150	10	4.2	210	15	4.7	315	24	5.2
10,001–20,000	39	1	2.2	65	3	3.0	120	7	3.7	150	10	4.3	240	17	4.8	340	26	5.4
20,001–50,000	39	1	2.2	80	4	3.2	120	7	3.7	165	11	4.4	260	19	5.0	380	30	5.7
50,001–100,000	39	1	2.2	95	5	3.3	130	8	4.0	180	12	4.4	270	20	5.1	380	30	5.7

n = sample size; c = acceptance number
"All" indicates that each piece in the lot is to be inspected
AOQL = Average Outgoing Quality Limit

Appendix 5

Double Sampling Tables for Stated Values of

Lot Tolerance Per Cent Defective (LTPD)

with Consumer's Risk of 0.10

Double Sampling Table for
Lot Tolerance Per Cent Defective (LTPD) = 0.5%

Lot Size	Process Average 0 to 0.005%						Process Average 0.006 to 0.050%						Process Average 0.051 to 0.100%					
	Trial 1		Trial 2			AOQL in %	Trial 1		Trial 2			AOQL in %	Trial 1		Trial 2			AOQL in %
	n_1	c_1	n_2	n_1+n_2	c_2		n_1	c_1	n_2	n_1+n_2	c_2		n_1	c_1	n_2	n_1+n_2	c_2	
1–180	All	0	–	–	–	0	All	0	–	–	–	0	All	0	–	–	–	0
181–210	180	0	–	–	–	0.02	180	0	–	–	–	0.02	180	0	–	–	–	0.02
211–250	210	0	–	–	–	0.03	210	0	–	–	–	0.03	210	0	–	–	–	0.03
251–300	240	0	–	–	–	0.03	240	0	–	–	–	0.03	240	0	–	–	–	0.03
301–400	275	0	–	–	–	0.04	275	0	–	–	–	0.04	275	0	–	–	–	0.04
401–450	290	0	–	–	–	0.04	290	0	–	–	–	0.04	290	0	–	–	–	0.04
451–500	340	0	110	450	1	0.04	340	0	110	450	1	0.04	340	0	110	450	1	0.04
501–550	350	0	130	480	1	0.05	350	0	130	480	1	0.05	350	0	130	480	1	0.05
551–600	360	0	150	510	1	0.05	360	0	150	510	1	0.05	360	0	150	510	1	0.05
601–800	400	0	185	585	1	0.06	400	0	185	585	1	0.06	400	0	185	585	1	0.06
801–1000	430	0	200	630	1	0.07	430	0	200	630	1	0.07	430	0	200	630	1	0.07
1001–2000	490	0	265	755	1	0.08	490	0	265	755	1	0.08	490	0	265	755	1	0.08
2001–3000	520	0	290	810	1	0.09	520	0	290	810	1	0.09	520	0	530	1050	2	0.10
3001–4000	530	0	310	840	1	0.09	530	0	570	1100	2	0.11	530	0	570	1100	2	0.11
4001–5000	540	0	305	845	1	0.09	540	0	580	1120	2	0.11	540	0	830	1370	3	0.12
5001–7000	545	0	315	860	1	0.10	545	0	615	1160	2	0.11	545	0	865	1410	3	0.12
7001–10,000	550	0	330	880	1	0.10	550	0	620	1170	2	0.12	550	0	1130	1680	4	0.14
10,001–20,000	555	0	345	900	1	0.10	555	0	925	1480	3	0.13	555	0	1185	1740	4	0.15
20,001–50,000	560	0	650	1210	2	0.12	560	0	940	1500	3	0.14	900	1	1400	2300	6	0.16
50,001–100,000	560	0	650	1210	2	0.12	560	0	1210	1770	4	0.15	905	1	1655	2560	7	0.17

Lot Size	Process Average 0.101 to 0.150%						Process Average 0.151 to 0.200%						Process Average 0.201 to 0.250%					
	Trial 1		Trial 2			AOQL in %	Trial 1		Trial 2			AOQL in %	Trial 1		Trial 2			AOQL in %
	n_1	c_1	n_2	n_1+n_2	c_2		n_1	c_1	n_2	n_1+n_2	c_2		n_1	c_1	n_2	n_1+n_2	c_2	
1–180	All	0	–	–	–	0	All	0	–	–	–	0	All	0	–	–	–	0
181–210	180	0	–	–	–	0.02	180	0	–	–	–	0.02	180	0	–	–	–	0.02
211–250	210	0	–	–	–	0.03	210	0	–	–	–	0.03	210	0	–	–	–	0.03
251–300	240	0	–	–	–	0.03	240	0	–	–	–	0.03	240	0	–	–	–	0.03
301–400	275	0	–	–	–	0.04	275	0	–	–	–	0.04	275	0	–	–	–	0.04
401–450	290	0	–	–	–	0.04	290	0	–	–	–	0.04	290	0	–	–	–	0.04
451–500	340	0	110	450	1	0.04	340	0	110	450	1	0.04	340	0	110	450	1	0.04
501–550	350	0	130	480	1	0.05	350	0	130	480	1	0.05	350	0	130	480	1	0.05
551–600	360	0	150	510	1	0.05	360	0	150	510	1	0.05	360	0	150	510	1	0.05
601–800	400	0	185	585	1	0.06	400	0	185	585	1	0.06	400	0	185	585	1	0.06
801–1000	430	0	200	630	1	0.07	430	0	200	630	1	0.07	430	0	200	630	1	0.07
1001–2000	490	0	500	990	2	0.09	490	0	500	990	2	0.09	490	0	500	990	2	0.09
2001–3000	520	0	530	1050	2	0.10	520	0	760	1280	3	0.11	520	0	980	1500	4	0.11
3001–4000	530	0	810	1340	3	0.11	530	0	1030	1560	4	0.12	840	1	1160	2000	6	0.13
4001–5000	540	0	1060	1600	4	0.13	845	1	1205	2050	6	0.14	845	1	1425	2270	7	0.14
5001–7000	545	0	1105	1650	4	0.13	860	1	1490	2350	7	0.15	860	1	1700	2560	8	0.16
7001–10,000	880	1	1300	2180	6	0.15	880	1	1770	2650	8	0.16	1170	2	2160	3330	11	0.17
10,001–20,000	900	1	1840	2740	8	0.18	1200	2	2250	3450	11	0.19	1740	4	2620	4360	15	0.21
20,001–50,000	1210	2	2330	3540	11	0.20	1500	3	2980	4480	15	0.22	2300	6	4240	6540	24	0.24
50,001–100,000	1210	2	2590	3800	12	0.21	1770	4	3690	5460	19	0.23	2560	7	5420	7980	30	0.26

Trial 1: n_1 = first sample size; c_1 = acceptance number for first sample
"All" indicates that each piece in the lot is to be inspected
Trial 2: n_2 = second sample size; c_2 = acceptance number for first and second samples combined
AOQL = Average Outgoing Quality Limit

Double Sampling Table for Lot Tolerance Per Cent Defective (LTPD) = 1.0%

Lot Size	Process Average 0 to 0.010%						Process Average 0.011 to 0.10%						Process Average 0.11 to 0.20%					
	Trial 1		Trial 2			AOQL in %	Trial 1		Trial 2			AOQL in %	Trial 1		Trial 2			AOQL in %
	n_1	c_1	n_2	n_1+n_2	c_2		n_1	c_1	n_2	n_1+n_2	c_2		n_1	c_1	n_2	n_1+n_2	c_2	
1–120	All	0	–	–	–	0	All	0	–	–	–	0	All	0	–	–	–	0
121–150	120	0	–	–	–	0.06	120	0	–	–	–	0.06	120	0	–	–	–	0.06
151–200	140	0	–	–	–	0.08	140	0	–	–	–	0.08	140	0	–	–	–	0.08
201–260	165	0	–	–	–	0.10	165	0	–	–	–	0.10	165	0	–	–	–	0.10
261–300	180	0	75	255	1	0.10	180	0	75	255	1	0.10	180	0	75	255	1	0.10
301–400	200	0	90	290	1	0.12	200	0	90	290	1	0.12	200	0	90	290	1	0.12
401–500	215	0	100	315	1	0.14	215	0	100	315	1	0.14	215	0	100	315	1	0.14
501–600	225	0	115	340	1	0.15	225	0	115	340	1	0.15	225	0	115	340	1	0.15
601–800	235	0	125	360	1	0.16	235	0	125	360	1	0.16	235	0	125	360	1	0.16
801–1000	245	0	135	380	1	0.17	245	0	135	380	1	0.17	245	0	250	495	2	0.19
1001–2000	265	0	155	420	1	0.18	265	0	155	420	1	0.18	265	0	285	550	2	0.21
2001–3000	270	0	160	430	1	0.19	270	0	300	570	2	0.22	270	0	420	690	3	0.25
3001–4000	275	0	160	435	1	0.19	275	0	305	580	2	0.22	275	0	435	710	3	0.25
4001–5000	275	0	165	440	1	0.19	275	0	310	585	2	0.23	275	0	565	840	4	0.28
5001–7000	275	0	170	445	1	0.20	275	0	315	590	2	0.23	275	0	580	855	4	0.29
7001–10,000	280	0	320	600	2	0.24	280	0	460	740	3	0.26	280	0	590	870	4	0.30
10,001–20,000	280	0	325	605	2	0.24	280	0	465	745	3	0.27	450	1	700	1150	6	0.33
20,001–50,000	280	0	325	605	2	0.25	280	0	605	885	4	0.30	450	1	830	1280	7	0.36
50,001–100,000	280	0	325	605	2	0.25	280	0	605	885	4	0.30	450	1	960	1410	8	0.38

Lot Size	Process Average 0.21 to 0.30%						Process Average 0.31 to 0.40%						Average Process 0.41 to 0.50%					
	Trial 1		Trial 2			AOQL in %	Trial 1		Trial 2			AOQL in %	Trial 1		Trial 2			AOQL in %
	n_1	c_1	n_2	n_1+n_2	c_2		n_1	c_1	n_2	n_1+n_2	c_2		n_1	c_1	n_2	n_1+n_2	c_2	
1–120	All	0	–	–	–	0	All	0	–	–	–	0	All	0	–	–	–	0
121–150	120	0	–	–	–	0.06	120	0	–	–	–	0.06	120	0	–	–	–	0.06
151–200	140	0	–	–	–	0.08	140	0	–	–	–	0.08	140	0	–	–	–	0.08
201–260	165	0	–	–	–	0.10	165	0	–	–	–	0.10	165	0	–	–	–	0.10
261–300	180	0	75	255	1	0.10	180	0	75	255	1	0.10	180	0	75	255	1	0.10
301–400	200	0	90	290	1	0.12	200	0	90	290	1	0.12	200	0	90	290	1	0.12
401–500	215	0	100	315	1	0.14	215	0	100	315	1	0.14	215	0	100	315	1	0.14
501–600	225	0	115	340	1	0.15	225	0	115	340	1	0.15	225	0	205	430	2	0.16
601–800	235	0	230	465	2	0.18	235	0	230	465	2	0.18	235	0	230	465	2	0.18
801–1000	245	0	250	495	2	0.19	245	0	250	495	2	0.19	245	0	250	495	2	0.19
1001–2000	265	0	405	670	3	0.23	265	0	515	780	4	0.24	265	0	515	780	4	0.24
2001–3000	270	0	545	815	4	0.26	430	1	620	1050	6	0.28	430	1	830	1260	8	0.30
3001–4000	435	1	645	1080	6	0.29	435	1	865	1300	8	0.30	580	2	940	1520	10	0.33
4001–5000	440	1	660	1100	6	0.30	440	1	1000	1440	9	0.33	585	2	1075	1660	11	0.35
5001–7000	445	1	785	1230	7	0.33	590	2	990	1580	10	0.36	730	3	1190	1920	13	0.38
7001–10,000	450	1	920	1370	8	0.35	600	2	1240	1840	12	0.39	870	4	1540	2410	17	0.41
10,001–20,000	605	2	1035	1640	10	0.39	745	3	1485	2230	15	0.43	1150	6	1990	3140	23	0.44
20,001–50,000	605	2	1295	1900	12	0.42	885	4	1845	2730	19	0.47	1280	7	2600	3880	29	0.52
50,001–100,000	605	2	1545	2150	14	0.44	885	4	2085	2970	21	0.49	1410	8	3280	4690	36	0.55

Trial 1: n_1 = first sample size; c_1 = acceptance number for first sample
"All" indicates that each piece in the lot is to be inspected
Trial 2: n_2 = second sample size; c_2 = acceptance number for first and second samples combined
AOQL = Average Outgoing Quality Limit

Double Sampling Table for
Lot Tolerance Per Cent Defective (LTPD) = 2.0%

Lot Size	Process Average 0 to 0.02% Trial 1 n_1	c_1	Trial 2 n_2	n_1+n_2	c_2	AOQL in %	Process Average 0.03 to 0.20% Trial 1 n_1	c_1	Trial 2 n_2	n_1+n_2	c_2	AOQL in %	Process Average 0.21 to 0.40% Trial 1 n_1	c_1	Trial 2 n_2	n_1+n_2	c_2	AOQL in %
1–75	All	0	–	–	–	0	All	0	–	–	–	0	All	0	–	–	–	0
76–100	70	0	–	–	–	0.16	70	0	–	–	–	0.16	70	0	–	–	–	0.16
101–200	85	0	–	–	–	0.25	85	0	–	–	–	0.25	85	0	–	–	–	0.25
201–300	115	0	50	165	1	0.29	115	0	50	165	1	0.29	115	0	50	165	1	0.29
301–400	120	0	60	180	1	0.32	120	0	60	180	1	0.32	120	0	60	180	1	0.32
401–500	125	0	65	190	1	0.33	125	0	65	190	1	0.33	125	0	120	245	2	0.37
501–600	125	0	70	195	1	0.34	125	0	70	195	1	0.34	125	0	130	255	2	0.39
601–800	130	0	75	205	1	0.35	130	0	75	205	1	0.35	130	0	135	265	2	0.41
801–1000	135	0	75	210	1	0.36	135	0	140	275	2	0.42	135	0	140	275	2	0.42
1001–2000	135	0	85	220	1	0.38	135	0	155	290	2	0.45	135	0	220	355	3	0.50
2001–3000	140	0	85	225	1	0.39	140	0	155	295	2	0.46	140	0	285	425	4	0.56
3001–4000	140	0	85	225	1	0.40	140	0	225	365	3	0.52	140	0	290	430	4	0.57
4001–5000	140	0	160	300	2	0.47	140	0	230	370	3	0.53	140	0	360	500	5	0.61
5001–7000	140	0	160	300	2	0.48	140	0	230	370	3	0.54	140	0	365	505	5	0.62
7001–10,000	140	0	160	300	2	0.48	140	0	235	375	3	0.54	225	1	350	575	6	0.66
10,001–20,000	140	0	165	305	2	0.49	140	0	235	375	3	0.54	225	1	415	640	7	0.71
20,001–50,000	140	0	165	305	2	0.49	140	0	305	445	4	0.59	225	1	480	705	8	0.75
50,001–100,000	140	0	165	305	2	0.49	140	0	305	445	4	0.60	225	1	545	770	9	0.78

Lot Size	Process Average 0.41 to 0.60% Trial 1 n_1	c_1	Trial 2 n_2	n_1+n_2	c_2	AOQL in %	Process Average 0.61 to 0.80% Trial 1 n_1	c_1	Trial 2 n_2	n_1+n_2	c_2	AOQL in %	Process Average 0.81 to 1.00% Trial 1 n_1	c_1	Trial 2 n_2	n_1+n_2	c_2	AOQL in %
1–75	All	0	–	–	–	0	All	0	–	–	–	0	All	0	–	–	–	0
76–100	70	0	–	–	–	0.16	70	0	–	–	–	0.16	70	0	–	–	–	0.16
101–200	85	0	–	–	–	0.25	85	0	–	–	–	0.25	85	0	–	–	–	0.25
201–300	115	0	50	165	1	0.29	115	0	50	165	1	0.29	115	0	50	165	1	0.29
301–400	120	0	115	235	2	0.34	120	0	115	235	2	0.34	120	0	115	235	2	0.34
401–500	125	0	120	245	2	0.37	125	0	120	245	2	0.37	125	0	120	245	2	0.37
501–600	125	0	130	255	2	0.39	125	0	185	310	3	0.41	125	0	185	310	3	0.41
601–800	130	0	195	325	3	0.44	130	0	250	380	4	0.45	130	0	250	380	4	0.45
801–1000	135	0	200	335	3	0.46	135	0	255	390	4	0.48	210	1	290	500	6	0.54
1001–2000	135	0	285	420	4	0.54	220	1	375	595	7	0.62	220	1	485	705	9	0.65
2001–3000	225	1	385	610	7	0.65	295	2	435	730	9	0.69	360	3	535	895	12	0.72
3001–4000	225	1	455	680	8	0.69	295	2	555	850	11	0.74	365	3	715	1080	15	0.77
4001–5000	225	1	460	685	8	0.70	300	2	620	920	12	0.77	435	4	775	1210	17	0.81
5001–7000	300	2	450	750	9	0.74	370	3	680	1050	14	0.82	505	5	935	1440	21	0.89
7001–10,000	300	2	520	820	10	0.77	375	3	735	1110	15	0.85	575	6	1055	1630	24	0.95
10,001–20,000	305	2	645	950	12	0.83	375	3	935	1310	18	0.92	640	7	1240	1880	28	1.0
20,001–50,000	305	2	715	1020	13	0.86	445	4	1045	1490	21	0.98	705	8	1635	2340	36	1.1
50,001–100,000	305	2	830	1135	15	0.90	510	5	1150	1660	24	1.0	770	9	1850	2620	41	1.2

Trial 1: n_1 = first sample size; c_1 = acceptance number for first sample
"All" indicates that each piece in the lot is to be inspected
Trial 2: n_2 = second sample size; c_2 = acceptance number for first and second samples combined
AOQL = Average Outgoing Quality Limit

Double Sampling Table for
Lot Tolerance Per Cent Defective (LTPD) = 3.0%

Lot Size	Process Average 0 to 0.03% Trial 1 n_1	c_1	Trial 2 n_2	n_1+n_2	c_2	AOQL in %	Process Average 0.04 to 0.30% Trial 1 n_1	c_1	Trial 2 n_2	n_1+n_2	c_2	AOQL in %	Process Average 0.31 to 0.60% Trial 1 n_1	c_1	Trial 2 n_2	n_1+n_2	c_2	AOQL in %
1–40	All	0	–	–	–	0	All	0	–	–	–	0	All	0	–	–	–	0
41–55	40	0	–	–	–	0.18	40	0	–	–	–	0.18	40	0	–	–	–	0.18
56–100	55	0	–	–	–	0.30	55	0	–	–	–	0.30	55	0	–	–	–	0.30
101–150	70	0	30	100	1	0.37	70	0	30	100	1	0.37	70	0	30	100	1	0.37
151–200	75	0	40	115	1	0.45	75	0	40	115	1	0.45	75	0	40	115	1	0.45
201–300	75	0	40	115	1	0.50	75	0	40	115	1	0.50	75	0	40	115	1	0.50
301–400	80	0	45	125	1	0.52	80	0	45	125	1	0.52	80	0	85	165	2	0.57
401–500	85	0	50	135	1	0.53	85	0	50	135	1	0.53	85	0	90	175	2	0.60
501–600	85	0	50	135	1	0.54	85	0	50	135	1	0.54	85	0	95	180	2	0.62
601–800	90	0	50	140	1	0.55	90	0	95	185	2	0.64	90	0	135	225	3	0.70
801–1000	90	0	55	145	1	0.56	90	0	100	190	2	0.66	90	0	140	230	3	0.72
1001–2000	90	0	60	150	1	0.58	90	0	105	195	2	0.70	90	0	190	280	4	0.84
2001–3000	90	0	60	150	1	0.59	90	0	155	245	3	0.80	90	0	200	290	4	0.86
3001–4000	95	0	105	200	2	0.72	95	0	150	245	3	0.80	95	0	235	330	5	0.92
4001–5000	95	0	105	200	2	0.73	95	0	155	250	3	0.81	150	1	230	380	6	0.98
5001–7000	95	0	105	200	2	0.73	95	0	155	250	3	0.81	150	1	230	380	6	1.0
7001–10,000	95	0	105	200	2	0.73	95	0	155	250	3	0.81	150	1	275	425	7	1.0
10,001–20,000	95	0	105	200	2	0.74	95	0	200	295	4	0.92	150	1	320	470	8	1.1
20,001–50,000	95	0	105	200	2	0.74	95	0	200	295	4	0.93	150	1	365	515	9	1.2
50,001–100,000	95	0	105	200	2	0.75	95	0	245	340	5	1.0	150	1	405	555	10	1.2

Lot Size	Process Average 0.61 to 0.90% Trial 1 n_1	c_1	Trial 2 n_2	n_1+n_2	c_2	AOQL in %	Process Average 0.91 to 1.20% Trial 1 n_1	c_1	Trial 2 n_2	n_1+n_2	c_2	AOQL in %	Process Average 1.21 to 1.50% Trial 1 n_1	c_1	Trial 2 n_2	n_1+n_2	c_2	AOQL in %
1–40	All	0	–	–	–	0	All	0	–	–	–	0	All	0	–	–	–	0
41–55	40	0	–	–	–	0.18	40	0	–	–	–	0.18	40	0	–	–	–	0.18
56–100	55	0	–	–	–	0.30	55	0	–	–	–	0.30	55	0	–	–	–	0.30
101–150	70	0	30	100	1	0.37	70	0	30	100	1	0.37	70	0	30	100	1	0.37
151–200	75	0	40	115	1	0.45	75	0	65	140	2	0.47	75	0	65	140	2	0.47
201–300	75	0	80	155	2	0.54	75	0	80	155	2	0.54	75	0	80	155	2	0.54
301–400	80	0	85	165	2	0.57	80	0	120	200	3	0.62	80	0	120	200	3	0.62
401–500	85	0	125	210	3	0.64	85	0	125	210	3	0.64	85	0	160	245	4	0.69
501–600	85	0	130	215	3	0.67	85	0	170	255	4	0.72	135	1	185	320	6	0.76
601–800	90	0	170	260	4	0.74	140	1	195	335	6	0.79	140	1	210	350	7	0.81
801–1000	90	0	180	270	4	0.77	145	1	235	380	7	0.85	145	1	270	415	8	0.86
1001–2000	150	1	210	360	6	0.90	150	1	325	475	9	1.0	195	2	350	545	11	1.1
2001–3000	150	1	300	450	8	1.0	200	2	365	565	11	1.1	290	4	470	760	16	1.2
3001–4000	150	1	350	500	9	1.1	245	3	405	650	13	1.2	330	5	545	875	19	1.2
4001–5000	200	2	340	540	10	1.2	250	3	445	695	14	1.2	380	6	620	1000	22	1.3
5001–7000	200	2	385	585	11	1.2	250	3	530	780	16	1.3	380	6	700	1080	24	1.4
7001–10,000	200	2	425	625	12	1.2	250	3	575	825	17	1.3	425	7	785	1210	27	1.5
10,001–20,000	200	2	475	675	13	1.3	295	4	655	950	20	1.4	470	8	900	1370	31	1.6
20,001–50,000	200	2	515	715	14	1.3	295	4	755	1050	23	1.5	515	9	1165	1680	39	1.7
50,001–100,000	200	2	555	755	15	1.3	340	5	840	1180	26	1.6	515	9	1315	1830	43	1.8

Trial 1: n_1 = first sample size; c_1 = acceptance number for first sample
"All" indicates that each piece in the lot is to be inspected
Trial 2: n_2 = second sample size; c_2 = acceptance number for first and second samples combined
AOQL = Average Outgoing Quality Limit

DOUBLE
SAMPLING

4.0%

LTPD

— 192 —

Appendix 5

Double Sampling Table for
Lot Tolerance Per Cent Defective (LTPD) = 4.0%

Lot Size	Process Average 0 to 0.04%						Process Average 0.05 to 0.40%						Process Average 0.41 to 0.80%					
	Trial 1		Trial 2			AOQL in %	Trial 1		Trial 2			AOQL in %	Trial 1		Trial 2			AOQL in %
	n_1	c_1	n_2	n_1+n_2	c_2		n_1	c_1	n_2	n_1+n_2	c_2		n_1	c_1	n_2	n_1+n_2	c_2	
1–35	All	0	–	–	–	0	All	0	–	–	–	0	All	0	–	–	–	0
36–50	34	0	–	–	–	0.35	34	0	–	–	–	0.35	34	0	–	–	–	0.35
51–75	40	0	–	–	–	0.43	40	0	–	–	–	0.43	40	0	–	–	–	0.43
76–100	50	0	25	75	1	0.46	50	0	25	75	1	0.46	50	0	25	75	1	0.46
101–150	55	0	30	85	1	0.55	55	0	30	85	1	0.55	55	0	30	85	1	0.55
151–200	60	0	30	90	1	0.64	60	0	30	90	1	0.64	60	0	30	90	1	0.64
201–300	60	0	35	95	1	0.70	60	0	35	95	1	0.70	60	0	65	125	2	0.75
301–400	65	0	35	100	1	0.71	65	0	35	100	1	0.71	65	0	65	130	2	0.80
401–500	65	0	40	105	1	0.73	65	0	70	135	2	0.83	65	0	70	135	2	0.83
501–600	65	0	40	105	1	0.74	65	0	75	140	2	0.85	65	0	100	165	3	0.93
601–800	65	0	40	105	1	0.75	65	0	75	140	2	0.87	65	0	110	175	3	0.97
801–1000	70	0	40	110	1	0.76	70	0	75	145	2	0.90	70	0	105	175	3	0.98
1001–2000	70	0	40	110	1	0.78	70	0	80	150	2	0.94	70	0	145	215	4	1.1
2001–3000	70	0	80	150	2	0.95	70	0	115	185	3	1.1	70	0	180	250	5	1.2
3001–4000	70	0	80	150	2	0.96	70	0	115	185	3	1.1	110	1	175	285	6	1.3
4001–5000	70	0	80	150	2	0.97	70	0	115	185	3	1.1	115	1	170	285	6	1.3
5001–7000	70	0	80	150	2	0.98	70	0	115	185	3	1.1	115	1	205	320	7	1.4
7001–10,000	70	0	80	150	2	0.98	70	0	150	220	4	1.2	115	1	205	320	7	1.4
10,001–20,000	70	0	80	150	2	0.98	70	0	150	220	4	1.2	115	1	235	350	8	1.5
20,001–50,000	70	0	80	150	2	0.99	70	0	150	220	4	1.2	115	1	270	385	9	1.6
50,001–100,000	70	0	80	150	2	0.99	70	0	185	255	5	1.3	115	1	300	415	10	1.7

Lot Size	Process Average 0.81 to 1.20%						Process Average 1.21 to 1.60%						Process Average 1.61 to 2.00%					
	Trial 1		Trial 2			AOQL in %	Trial 1		Trial 2			AOQL in %	Trial 1		Trial 2			AOQL in %
	n_1	c_1	n_2	n_1+n_2	c_2		n_1	c_1	n_2	n_1+n_2	c_2		n_1	c_1	n_2	n_1+n_2	c_2	
1–35	All	0	–	–	–	0	All	0	–	–	–	0	All	0	–	–	–	0
36–50	34	0	–	–	–	0.35	34	0	–	–	–	0.35	34	0	–	–	–	0.35
51–75	40	0	–	–	–	0.43	40	0	–	–	–	0.43	40	0	–	–	–	0.43
76–100	50	0	25	75	1	0.46	50	0	25	75	1	0.46	50	0	25	75	1	0.46
101–150	55	0	30	85	1	0.55	55	0	30	85	1	0.55	55	0	30	85	1	0.55
151–200	60	0	55	115	2	0.68	60	0	55	115	2	0.68	60	0	55	115	2	0.68
201–300	60	0	65	125	2	0.75	60	0	90	150	3	0.84	60	0	90	150	3	0.84
301–400	65	0	95	160	3	0.86	65	0	95	160	3	0.86	65	0	120	185	4	0.92
401–500	65	0	100	165	3	0.92	65	0	130	195	4	0.96	105	1	140	245	6	1.0
501–600	65	0	135	200	4	1.0	105	1	145	250	6	1.1	105	1	175	280	7	1.1
601–800	65	0	140	205	4	1.0	105	1	185	290	7	1.2	105	1	210	315	8	1.2
801–1000	110	1	155	265	6	1.2	110	1	210	320	8	1.2	145	2	230	375	10	1.3
1001–2000	110	1	195	305	7	1.3	150	2	240	390	10	1.5	180	3	295	475	13	1.6
2001–3000	110	1	260	370	9	1.4	185	3	305	490	13	1.6	220	4	410	630	18	1.7
3001–4000	150	2	255	405	10	1.5	185	3	340	525	14	1.6	285	6	465	750	22	1.8
4001–5000	150	2	285	435	11	1.6	185	3	395	580	16	1.7	285	6	520	805	24	1.9
5001–7000	150	2	320	470	12	1.6	185	3	435	620	17	1.7	320	7	585	905	27	2.0
7001–10,000	150	2	325	475	12	1.7	220	4	460	680	19	1.9	320	7	645	965	29	2.1
10,001–20,000	150	2	355	505	13	1.7	220	4	495	715	20	1.9	350	8	790	1140	35	2.2
20,001–50,000	150	2	420	570	15	1.7	255	5	575	830	24	2.0	385	9	895	1280	40	2.3
50,001–100,000	150	2	450	600	16	1.8	255	5	665	920	27	2.1	415	10	985	1400	44	2.4

Trial 1: n_1 = first sample size; c_1 = acceptance number for first sample
"All" indicates that each piece in the lot is to be inspected
Trial 2: n_2 = second sample size; c_2 = acceptance number for first and second samples combined
AOQL = Average Outgoing Quality Limit

Double Sampling Table for
Lot Tolerance Per Cent Defective (LTPD) = 5.0%

Lot Size	Process Average 0 to 0.05%						Process Average 0.06 to 0.50%						Process Average 0.51 to 1.00%					
	Trial 1		Trial 2			AOQL in %	Trial 1		Trial 2			AOQL in %	Trial 1		Trial 2			AOQL in %
	n_1	c_1	n_2	n_1+n_2	c_2		n_1	c_1	n_2	n_1+n_2	c_2		n_1	c_1	n_2	n_1+n_2	c_2	
1–30	All	0	–	–	–	0	All	0	–	–	–	0	All	0	–	–	–	0
31–50	30	0	–	–	–	0.49	30	0	–	–	–	0.49	30	0	–	–	–	0.49
51–75	38	0	–	–	–	0.59	38	0	–	–	–	0.59	38	0	–	–	–	0.59
76–100	44	0	21	65	1	0.64	44	0	21	65	1	0.64	44	0	21	65	1	0.64
101–200	49	0	26	75	1	0.84	49	0	26	75	1	0.84	49	0	26	75	1	0.84
201–300	50	0	30	80	1	0.91	50	0	30	80	1	0.91	50	0	55	105	2	1.0
301–400	55	0	30	85	1	0.92	55	0	55	110	2	1.1	55	0	55	110	2	1.1
401–500	55	0	30	85	1	0.93	55	0	55	110	2	1.1	55	0	80	135	3	1.2
501–600	55	0	30	85	1	0.94	55	0	60	115	2	1.1	55	0	85	140	3	1.2
601–800	55	0	35	90	1	0.95	55	0	65	120	2	1.1	55	0	85	140	3	1.3
801–1000	55	0	35	90	1	0.96	55	0	65	120	2	1.1	55	0	115	170	4	1.4
1001–2000	55	0	35	90	1	0.98	55	0	95	150	3	1.3	55	0	120	175	4	1.4
2001–3000	55	0	65	120	2	1.2	55	0	95	150	3	1.3	55	0	150	205	5	1.5
3001–4000	55	0	65	120	2	1.2	55	0	95	150	3	1.3	90	1	140	230	6	1.6
4001–5000	55	0	65	120	2	1.2	55	0	95	150	3	1.4	90	1	165	255	7	1.8
5001–7000	55	0	65	120	2	1.2	55	0	95	150	3	1.4	90	1	165	255	7	1.8
7001–10,000	55	0	65	120	2	1.2	55	0	120	175	4	1.5	90	1	190	280	8	1.9
10,001–20,000	55	0	65	120	2	1.2	55	0	120	175	4	1.5	90	1	190	280	8	1.9
20,001–50,000	55	0	65	120	2	1.2	55	0	150	205	5	1.7	90	1	215	305	9	2.0
50,001–100,000	55	0	65	120	2	1.2	55	0	150	205	5	1.7	90	1	240	330	10	2.1

Lot Size	Process Average 1.01 to 1.50%						Process Average 1.51 to 2.00%						Process Average 2.01 to 2.50%					
	Trial 1		Trial 2			AOQL in %	Trial 1		Trial 2			AOQL in %	Trial 1		Trial 2			AOQL in %
	n_1	c_1	n_2	n_1+n_2	c_2		n_1	c_1	n_2	n_1+n_2	c_2		n_1	c_1	n_2	n_1+n_2	c_2	
1–30	All	0	–	–	–	0	All	0	–	–	–	0	All	0	–	–	–	0
31–50	30	0	–	–	–	0.49	30	0	–	–	–	0.49	30	0	–	–	–	0.49
51–75	38	0	–	–	–	0.59	38	0	–	–	–	0.59	38	0	–	–	–	0.59
76–100	44	0	21	65	1	0.64	44	0	21	65	1	0.64	44	0	21	65	1	0.64
101–200	49	0	51	100	2	0.91	49	0	51	100	2	0.91	49	0	51	100	2	0.91
201–300	50	0	55	105	2	1.0	50	0	80	130	3	1.1	50	0	100	150	4	1.1
301–400	55	0	80	135	3	1.1	55	0	100	155	4	1.2	85	1	105	190	6	1.3
401–500	55	0	105	160	4	1.3	85	1	120	205	6	1.4	85	1	140	225	7	1.4
501–600	55	0	110	165	4	1.3	85	1	145	230	7	1.4	85	1	165	250	8	1.5
601–800	90	1	125	215	6	1.5	90	1	170	260	8	1.5	120	2	185	305	10	1.6
801–1000	90	1	150	240	7	1.5	90	1	200	290	9	1.6	120	2	210	330	11	1.7
1001–2000	90	1	185	275	8	1.7	120	2	225	345	11	1.9	175	4	260	435	15	2.0
2001–3000	120	2	180	300	9	1.9	150	3	270	420	14	2.1	205	5	375	580	21	2.3
3001–4000	120	2	210	330	10	2.0	150	3	295	445	15	2.3	230	6	420	650	24	2.4
4001–5000	120	2	255	375	12	2.1	150	3	345	495	17	2.3	255	7	445	700	26	2.5
5001–7000	120	2	260	380	12	2.1	150	3	370	520	18	2.3	255	7	495	750	28	2.6
7001–10,000	120	2	285	405	13	2.1	175	4	370	545	19	2.4	280	8	540	820	31	2.7
10,001–20,000	120	2	310	430	14	2.2	175	4	420	595	21	2.4	280	8	660	940	36	2.8
20,001–50,000	120	2	335	455	15	2.2	205	5	485	690	25	2.5	305	9	745	1050	41	2.9
50,001–100,000	120	2	360	480	16	2.3	205	5	555	760	28	2.6	330	10	810	1140	45	3.0

Trial 1: n_1 = first sample size; c_1 = acceptance number for first sample
"All" indicates that each piece in the lot is to be inspected
Trial 2: n_2 = second sample size; c_2 = acceptance number for first and second samples combined
AOQL = Average Outgoing Quality Limit

DOUBLE
SAMPLING

7.0%

LTPD

— 194 —

Appendix 5

Double Sampling Table for Lot Tolerance Per Cent Defective (LTPD) = 7.0%

Lot Size	Process Average 0 to 0.07%						Process Average 0.08 to 0.70%						Process Average 0.71 to 1.40%					
	Trial 1		Trial 2			AOQL in %	Trial 1		Trial 2			AOQL in %	Trial 1		Trial 2			AOQL in %
	n_1	c_1	n_2	n_1+n_2	c_2		n_1	c_1	n_2	n_1+n_2	c_2		n_1	c_1	n_2	n_1+n_2	c_2	
1–25	All	0	–	–	–	0	All	0	–	–	–	0	All	0	–	–	–	0
26–50	24	0	–	–	–	0.80	24	0	–	–	–	0.80	24	0	–	–	–	0.80
51–75	31	0	15	46	1	0.90	31	0	15	46	1	0.90	31	0	15	46	1	0.90
76–110	34	0	16	50	1	1.1	34	0	16	50	1	1.1	34	0	16	50	1	1.1
111–200	36	0	19	55	1	1.2	36	0	19	55	1	1.2	36	0	39	75	2	1.4
201–300	37	0	23	60	1	1.3	37	0	23	60	1	1.3	37	0	38	75	2	1.5
301–400	38	0	22	60	1	1.3	38	0	42	80	2	1.5	38	0	57	95	3	1.7
401–500	39	0	21	60	1	1.3	39	0	41	80	2	1.5	39	0	61	100	3	1.7
501–600	39	0	26	65	1	1.3	39	0	46	85	2	1.6	39	0	61	100	3	1.7
601–800	39	0	26	65	1	1.4	39	0	46	85	2	1.6	39	0	81	120	4	1.9
801–1000	39	0	26	65	1	1.4	39	0	46	85	2	1.6	39	0	86	125	4	2.0
1001–2000	40	0	45	85	2	1.7	40	0	65	105	3	1.9	40	0	105	145	5	2.2
2001–3000	40	0	45	85	2	1.7	40	0	65	105	3	1.9	65	1	100	165	6	2.3
3001–4000	40	0	45	85	2	1.7	40	0	65	105	3	1.9	65	1	115	180	7	2.4
4001–5000	40	0	45	85	2	1.7	40	0	85	125	4	2.1	65	1	115	180	7	2.5
5001–7000	40	0	45	85	2	1.7	40	0	85	125	4	2.1	65	1	135	200	8	2.5
7001–10,000	40	0	45	85	2	1.7	40	0	85	125	4	2.1	65	1	135	200	8	2.6
10,001–20,000	40	0	45	85	2	1.7	40	0	85	125	4	2.1	65	1	155	220	9	2.7
20,001–50,000	40	0	45	85	2	1.7	40	0	105	145	5	2.3	65	1	170	235	10	2.8
50,001–100,000	40	0	45	85	2	1.7	40	0	105	145	5	2.3	65	1	170	235	10	2.9

Lot Size	Process Average 1.41 to 2.10%						Process Average 2.11 to 2.80%						Process Average 2.81 to 3.50%					
	Trial 1		Trial 2			AOQL in %	Trial 1		Trial 2			AOQL in %	Trial 1		Trial 2			AOQL in %
	n_1	c_1	n_2	n_1+n_2	c_2		n_1	c_1	n_2	n_1+n_2	c_2		n_1	c_1	n_2	n_1+n_2	c_2	
1–25	All	0	–	–	–	0	All	0	–	–	–	0	All	0	–	–	–	0
26–50	24	0	–	–	–	0.80	24	0	–	–	–	0.80	24	0	–	–	–	0.80
51–75	31	0	15	46	1	0.90	31	0	15	46	1	0.90	31	0	15	46	1	0.90
76–110	34	0	31	65	2	1.2	34	0	31	65	2	1.2	34	0	31	65	2	1.2
111–200	36	0	54	90	3	1.5	36	0	54	90	3	1.5	36	0	69	105	4	1.5
201–300	37	0	58	95	3	1.6	37	0	73	110	4	1.7	60	1	80	140	6	1.9
301–400	38	0	77	115	4	1.8	60	1	85	145	6	1.9	60	1	100	160	7	2.0
401–500	39	0	76	115	4	1.8	60	1	105	165	7	2.1	60	1	135	195	9	2.2
501–600	65	1	90	155	6	2.1	65	1	120	185	8	2.2	85	2	130	215	10	2.3
601–800	65	1	105	170	7	2.2	65	1	140	205	9	2.3	85	2	165	250	12	2.5
801–1000	65	1	110	175	7	2.2	85	2	140	225	10	2.5	105	3	180	285	14	2.7
1001–2000	65	1	150	215	9	2.5	105	3	175	280	13	2.8	145	5	230	375	19	3.1
2001–3000	85	2	165	250	11	2.7	105	3	210	315	15	3.0	165	6	300	465	24	3.3
3001–4000	85	2	185	270	12	2.8	105	3	250	355	17	3.1	180	7	335	515	27	3.4
4001–5000	85	2	185	270	12	2.9	125	4	245	370	18	3.2	180	7	370	550	29	3.6
5001–7000	85	2	205	290	13	3.0	125	4	265	390	19	3.3	200	8	385	585	31	3.8
7001–10,000	85	2	205	290	13	3.1	125	4	300	425	21	3.4	200	8	450	650	35	3.9
10,001–20,000	85	2	220	305	14	3.2	125	4	335	460	23	3.5	220	9	485	705	38	4.0
20,001–50,000	85	2	240	325	15	3.3	145	5	360	505	26	3.5	220	9	565	785	43	4.1
50,001–100,000	105	3	270	375	18	3.4	165	6	390	555	29	3.7	235	10	610	845	47	4.2

Trial 1: n_1 = first sample size; c_1 = acceptance number for first sample
"All" indicates that each piece in the lot is to be inspected
Trial 2: n_2 = second sample size; c_2 = acceptance number for first and second samples combined
AOQL = Average Outgoing Quality Limit

Double Sampling Table for
Lot Tolerance Per Cent Defective (LTPD) = 10.0%

Lot Size	Process Average 0 to 0.10%						Process Average 0.11 to 1.00%						Process Average 1.01 to 2.00%					
	Trial 1		Trial 2			AOQL in %	Trial 1		Trial 2			AOQL in %	Trial 1		Trial 2			AOQL in %
	n_1	c_1	n_2	n_1+n_2	c_2		n_1	c_1	n_2	n_1+n_2	c_2		n_1	c_1	n_2	n_1+n_2	c_2	
1–20	All	0	–	–	–	0	All	0	–	–	–	0	All	0	–	–	–	0
21–50	17	0	–	–	–	1.3	17	0	–	–	–	1.3	17	0	–	–	–	1.3
51–100	25	0	13	38	1	1.6	25	0	13	38	1	1.6	25	0	13	38	1	1.6
101–200	27	0	15	42	1	1.8	27	0	15	42	1	1.8	27	0	28	55	2	2.1
201–300	27	0	16	43	1	1.9	27	0	30	57	2	2.2	27	0	43	70	3	2.4
301–400	27	0	17	44	1	1.9	27	0	33	60	2	2.2	27	0	43	70	3	2.5
401–500	28	0	16	44	1	1.9	28	0	32	60	2	2.3	28	0	57	85	4	2.7
501–600	28	0	17	45	1	1.9	28	0	32	60	2	2.3	28	0	57	85	4	2.8
601–800	28	0	17	45	1	2.0	28	0	47	75	3	2.6	28	0	57	85	4	2.9
801–1000	28	0	32	60	2	2.3	28	0	47	75	3	2.6	28	0	72	100	5	3.0
1001–2000	28	0	32	60	2	2.4	28	0	47	75	3	2.7	45	1	70	115	6	3.3
2001–3000	28	0	32	60	2	2.4	28	0	47	75	3	2.7	45	1	85	130	7	3.5
3001–4000	28	0	32	60	2	2.4	28	0	62	90	4	2.9	45	1	85	130	7	3.5
4001–5000	28	0	32	60	2	2.4	28	0	62	90	4	3.0	45	1	95	140	8	3.7
5001–7000	28	0	32	60	2	2.4	28	0	62	90	4	3.0	45	1	95	140	8	3.8
7001–10,000	28	0	32	60	2	2.5	28	0	62	90	4	3.0	45	1	95	140	8	3.8
10,001–20,000	28	0	32	60	2	2.5	28	0	62	90	4	3.0	45	1	110	155	9	3.9
20,001–50,000	28	0	32	60	2	2.5	28	0	72	100	5	3.3	45	1	120	165	10	3.9
50,001–100,000	28	0	32	60	2	2.5	28	0	72	100	5	3.3	45	1	135	180	11	4.2

Lot Size	Process Average 2.01 to 3.00%						Process Average 3.01 to 4.00%						Process Average 4.01 to 5.00%					
	Trial 1		Trial 2			AOQL in %	Trial 1		Trial 2			AOQL in %	Trial 1		Trial 2			AOQL in %
	n_1	c_1	n_2	n_1+n_2	c_2		n_1	c_1	n_2	n_1+n_2	c_2		n_1	c_1	n_2	n_1+n_2	c_2	
1–20	All	0	–	–	–	0	All	0	–	–	–	0	All	0	–	–	–	0
21–50	17	0	–	–	–	1.3	17	0	–	–	–	1.3	17	0	–	–	–	1.3
51–100	25	0	24	49	2	1.8	25	0	24	49	2	1.8	25	0	24	49	2	1.8
101–200	27	0	38	65	3	2.3	27	0	53	80	4	2.4	27	0	53	80	4	2.4
201–300	27	0	53	80	4	2.7	43	1	62	105	6	2.8	43	1	82	125	8	3.0
301–400	44	1	66	110	6	2.9	44	1	86	130	8	3.1	60	2	90	150	10	3.2
401–500	44	1	76	120	7	3.1	44	1	101	145	9	3.3	60	2	105	165	11	3.4
501–600	45	1	75	120	7	3.3	60	2	100	160	10	3.4	75	3	115	190	13	3.6
601–800	45	1	90	135	8	3.5	60	2	110	170	11	3.7	75	3	140	215	15	3.9
801–1000	45	1	90	135	8	3.5	60	2	125	185	12	3.9	90	4	150	240	17	4.1
1001–2000	60	2	105	165	10	3.9	75	3	150	225	15	4.3	115	6	200	315	23	4.8
2001–3000	60	2	130	190	12	4.1	75	3	175	250	17	4.4	130	7	235	365	27	5.0
3001–4000	60	2	130	190	12	4.2	90	4	170	260	18	4.6	130	7	255	385	29	5.1
4001–5000	60	2	140	200	13	4.3	90	4	180	270	19	4.7	140	8	270	410	31	5.2
5001–7000	60	2	140	200	13	4.4	90	4	205	295	21	4.9	140	8	315	455	35	5.3
7001–10,000	60	2	155	215	14	4.4	90	4	220	310	22	5.0	140	8	340	480	37	5.4
10,001–20,000	60	2	165	225	15	4.4	100	5	230	330	24	5.1	155	9	370	525	41	5.6
20,001–50,000	75	3	165	240	16	4.5	100	5	280	380	28	5.2	165	10	405	570	45	5.7
50,001–100,000	75	3	200	275	19	4.8	115	6	285	400	30	5.3	165	10	440	605	48	6.2

Trial 1: n_1 = first sample size; c_1 = acceptance number for first sample
"All" indicates that each piece in the lot is to be inspected
Trial 2: n_2 = second sample size; c_2 = acceptance number for first and second samples combined
AOQL = Average Outgoing Quality Limit

Appendix 6

Single Sampling Tables for Stated Values of Average Outgoing Quality Limit (AOQL)

0.1% AOQL

Single Sampling Table for Average Outgoing Quality Limit (AOQL) = 0.1%

Lot Size	Process Average 0 to 0.002%			Process Average 0.003 to 0.020%			Process Average 0.021 to 0.040%			Process Average 0.041 to 0.060%			Process Average 0.061 to 0.080%			Process Average 0.081 to 0.100%		
	n	c	p_t %	n	c	p_t %	n	c	p_t %	n	c	p_t %	n	c	p_t %	n	c	p_t %
1–75	All	0	–	All	0	–	All	0	–	All	0	–	All	0	–	All	0	–
76–95	75	0	1.5	75	0	1.5	75	0	1.5	75	0	1.5	75	0	1.5	75	0	1.5
96–130	95	0	1.4	95	0	1.4	95	0	1.4	95	0	1.4	95	0	1.4	95	0	1.4
131–200	130	0	1.2	130	0	1.2	130	0	1.2	130	0	1.2	130	0	1.2	130	0	1.2
201–300	165	0	1.1	165	0	1.1	165	0	1.1	165	0	1.1	165	0	1.1	165	0	1.1
301–400	190	0	0.96	190	0	0.96	190	0	0.96	190	0	0.96	190	0	0.96	190	0	0.96
401–500	210	0	0.91	210	0	0.91	210	0	0.91	210	0	0.91	210	0	0.91	210	0	0.91
501–600	230	0	0.86	230	0	0.86	230	0	0.86	230	0	0.86	230	0	0.86	230	0	0.86
601–800	250	0	0.81	250	0	0.81	250	0	0.81	250	0	0.81	250	0	0.81	250	0	0.81
801–1000	270	0	0.76	270	0	0.76	270	0	0.76	270	0	0.76	270	0	0.76	270	0	0.76
1001–2000	310	0	0.71	310	0	0.71	310	0	0.71	310	0	0.71	310	0	0.71	310	0	0.71
2001–3000	330	0	0.67	330	0	0.67	330	0	0.67	330	0	0.67	330	0	0.67	655	1	0.64
3001–4000	340	0	0.64	340	0	0.64	340	0	0.64	695	1	0.59	695	1	0.59	695	1	0.59
4001–5000	345	0	0.62	345	0	0.62	345	0	0.62	720	1	0.54	720	1	0.54	720	1	0.54
5001–7000	350	0	0.61	350	0	0.61	750	1	0.51	750	1	0.51	750	1	0.51	750	1	0.51
7001–10,000	355	0	0.60	355	0	0.60	775	1	0.49	775	1	0.49	775	1	0.49	1210	2	0.44
10,001–20,000	360	0	0.59	810	1	0.48	810	1	0.48	1280	2	0.42	1280	2	0.42	1770	3	0.38
20,001–50,000	365	0	0.58	830	1	0.47	1330	2	0.41	1870	3	0.37	2420	4	0.34	2980	5	0.33
50,001–100,000	365	0	0.58	835	1	0.46	1350	2	0.40	2480	4	0.33	3070	5	0.32	4270	7	0.30

0.25%

Single Sampling Table for Average Outgoing Quality Limit (AQQL) = 0.25%

Lot Size	Process Average 0 to 0.005%			Process Average 0.006 to 0.050%			Process Average 0.051 to 0.100%			Process Average 0.101 to 0.150%			Process Average 0.151 to 0.200%			Process Average 0.201 to 0.250%		
	n	c	p_t %	n	c	p_t %	n	c	p_t %	n	c	p_t %	n	c	p_t %	n	c	p_t %
1–60	All	0	–	All	0	–	All	0	–	All	0	–	All	0	–	All	0	–
61–100	60	0	2.5	60	0	2.5	60	0	2.5	60	0	2.5	60	0	2.5	60	0	2.5
101–200	85	0	2.1	85	0	2.1	85	0	2.1	85	0	2.1	85	0	2.1	85	0	2.1
201–300	100	0	1.9	100	0	1.9	100	0	1.9	100	0	1.9	100	0	1.9	100	0	1.9
301–400	110	0	1.8	110	0	1.8	110	0	1.8	110	0	1.8	110	0	1.8	110	0	1.8
401–500	115	0	1.8	115	0	1.8	115	0	1.8	115	0	1.8	115	0	1.8	115	0	1.8
501–600	120	0	1.7	120	0	1.7	120	0	1.7	120	0	1.7	120	0	1.7	120	0	1.7
601–800	125	0	1.7	125	0	1.7	125	0	1.7	125	0	1.7	125	0	1.7	125	0	1.7
801–1000	130	0	1.7	130	0	1.7	130	0	1.7	130	0	1.7	130	0	1.7	250	1	1.4
1001–2000	135	0	1.6	135	0	1.6	135	0	1.6	290	1	1.3	290	1	1.3	290	1	1.3
2001–3000	140	0	1.6	140	0	1.6	300	1	1.3	300	1	1.3	300	1	1.3	300	1	1.3
3001–4000	140	0	1.6	140	0	1.6	310	1	1.3	310	1	1.3	310	1	1.3	485	2	1.1
4001–5000	145	0	1.6	145	0	1.6	315	1	1.2	315	1	1.2	495	2	1.1	495	2	1.1
5001–7000	145	0	1.6	320	1	1.2	320	1	1.2	510	2	1.0	510	2	1.0	700	3	0.94
7001–10,000	145	0	1.6	325	1	1.2	325	1	1.2	520	2	1.0	720	3	0.91	720	3	0.91
10,001–20,000	145	0	1.6	330	1	1.2	535	2	1.0	750	3	0.89	970	4	0.81	1190	5	0.75
20,001–50,000	145	0	1.6	335	1	1.2	545	2	1.0	995	4	0.80	1240	5	0.74	1980	8	0.66
50,001–100,000	335	1	1.2	545	2	1.0	775	3	0.87	1250	5	0.73	1750	7	0.67	2810	11	0.62

n = sample size; c = acceptance number
"All" indicates that each piece in the lot is to be inspected
p_t = lot tolerance per cent defective with a Consumer's Risk (P_C) of 0.10

Single Sampling Table for
Average Outgoing Quality Limit (AOQL) = 0.5%

Lot Size	Process Average 0 to 0.010%			Process Average 0.011 to 0.10%			Process Average 0.11 to 0.20%			Process Average 0.21 to 0.30%			Process Average 0.31 to 0.40%			Process Average 0.41 to 0.50%		
	n	c	p_t %	n	c	p_t %	n	c	p_t %	n	c	p_t %	n	c	p_t %	n	c	p_t %
1–30	All	0	–	All	0	–	All	0	–	All	0	–	All	0	–	All	0	–
31–50	30	0	5.0	30	0	5.0	30	0	5.0	30	0	5.0	30	0	5.0	30	0	5.0
51–100	42	0	4.2	42	0	4.2	42	0	4.2	42	0	4.2	42	0	4.2	42	0	4.2
101–200	55	0	3.6	55	0	3.6	55	0	3.6	55	0	3.6	55	0	3.6	55	0	3.6
201–300	60	0	3.4	60	0	3.4	60	0	3.4	60	0	3.4	60	0	3.4	60	0	3.4
301–400	60	0	3.5	60	0	3.5	60	0	3.5	60	0	3.5	60	0	3.5	60	0	3.5
401–500	65	0	3.3	65	0	3.3	65	0	3.3	65	0	3.3	65	0	3.3	125	1	2.9
501–600	65	0	3.3	65	0	3.3	65	0	3.3	65	0	3.3	130	1	2.7	130	1	2.7
601–800	65	0	3.4	65	0	3.4	65	0	3.4	140	1	2.6	140	1	2.6	140	1	2.6
801–1000	70	0	3.2	70	0	3.2	70	0	3.2	145	1	2.6	145	1	2.6	145	1	2.6
1001–2000	70	0	3.2	70	0	3.2	155	1	2.5	155	1	2.5	155	1	2.5	240	2	2.2
2001–3000	70	0	3.3	70	0	3.3	160	1	2.4	160	1	2.4	250	2	2.1	250	2	2.1
3001–4000	70	0	3.3	160	1	2.4	160	1	2.4	255	2	2.1	255	2	2.1	355	3	1.9
4001–5000	75	0	3.0	165	1	2.4	165	1	2.4	260	2	2.0	360	3	1.9	460	4	1.7
5001–7000	75	0	3.0	165	1	2.4	265	2	2.0	265	2	2.0	370	3	1.8	475	4	1.7
7001–10,000	75	0	3.1	165	1	2.4	265	2	2.0	375	3	1.8	485	4	1.7	595	5	1.6
10,001–20,000	75	0	3.1	165	1	2.4	270	2	1.9	380	3	1.7	615	5	1.5	855	7	1.4
20,001–50,000	170	1	2.3	275	2	1.9	390	3	1.7	625	5	1.5	875	7	1.3	1410	11	1.2
50,001–100,000	170	1	2.3	275	2	1.9	510	4	1.6	755	6	1.4	1290	10	1.2	2130	16	1.1

Single Sampling Table for
Average Outgoing Quality Limit (AOQL) = 0.75%

Lot Size	Process Average 0 to 0.015%			Process Average 0.016 to 0.15%			Process Average 0.16 to 0.30%			Process Average 0.31 to 0.45%			Process Average 0.46 to 0.60%			Process Average 0.61 to 0.75%		
	n	c	p_t %	n	c	p_t %	n	c	p_t %	n	c	p_t %	n	c	p_t %	n	c	p_t %
1–25	All	0	–	All	0	–	All	0	–	All	0	–	All	0	–	All	0	–
26–50	25	0	6.4	25	0	6.4	25	0	6.4	25	0	6.4	25	0	6.4	25	0	6.4
51–100	33	0	5.6	33	0	5.6	33	0	5.6	33	0	5.6	33	0	5.6	33	0	5.6
101–200	39	0	5.2	39	0	5.2	39	0	5.2	39	0	5.2	39	0	5.2	39	0	5.2
201–300	42	0	5.0	42	0	5.0	42	0	5.0	42	0	5.0	42	0	5.0	42	0	5.0
301–400	44	0	4.9	44	0	4.9	44	0	4.9	44	0	4.9	90	1	4.0	90	1	4.0
401–500	45	0	4.8	45	0	4.8	45	0	4.8	90	1	4.1	90	1	4.1	90	1	4.1
501–600	45	0	4.9	45	0	4.9	45	0	4.9	95	1	3.9	95	1	3.9	95	1	3.9
601–800	46	0	4.9	46	0	4.9	100	1	3.8	100	1	3.8	100	1	3.8	100	1	3.8
801–1000	47	0	4.8	47	0	4.8	100	1	3.8	100	1	3.8	100	1	3.8	155	2	3.2
1001–2000	48	0	4.7	48	0	4.7	105	1	3.7	105	1	3.7	170	2	3.1	170	2	3.1
2001–3000	48	0	4.7	110	1	3.5	110	1	3.5	170	2	3.1	170	2	3.1	240	3	2.8
3001–4000	48	0	4.7	110	1	3.5	110	1	3.5	175	2	3.1	245	3	2.7	315	4	2.5
4001–5000	49	0	4.6	110	1	3.6	175	2	3.1	175	2	3.1	245	3	2.7	320	4	2.5
5001–7000	49	0	4.6	110	1	3.6	180	2	3.0	250	3	2.7	325	4	2.5	400	5	2.3
7001–10,000	49	0	4.6	110	1	3.7	180	2	3.0	255	3	2.6	405	5	2.3	560	7	2.1
10,001–20,000	49	0	4.6	110	1	3.7	255	3	2.6	335	4	2.4	495	6	2.1	750	9	1.9
20,001–50,000	110	1	3.7	180	2	3.0	260	3	2.6	420	5	2.2	675	8	1.9	1130	13	1.6
50,001–100,000	110	1	3.7	185	2	2.9	335	4	2.4	590	7	2.0	955	11	1.7	1720	19	1.5

n = sample size; c = acceptance number
"All" indicates that each piece in the lot is to be inspected
p_t = lot tolerance per cent defective with a Consumer's Risk (P_C) of 0.10

1.0%
AOQL

Single Sampling Table for
Average Outgoing Quality Limit (AOQL) = 1.0%

Lot Size	Process Average 0 to 0.02%			Process Average 0.03 to 0.20%			Process Average 0.21 to 0.40%			Process Average 0.41 to 0.60%			Process Average 0.61 to 0.80%			Process Average 0.81 to 1.00%.		
	n	c	p_t %	n	c	p_t %	n	c	p_t %	n	c	p_t %	n	c	p_t %	n	c	p_t %
1–25	All	0	–	All	0	–	All	0	–	All	0	–	All	0	–	All	0	–
26–50	22	0	7.7	22	0	7.7	22	0	7.7	22	0	7.7	22	0	7.7	22	0	7.7
51–100	27	0	7.1	27	0	7.1	27	0	7.1	27	0	7.1	27	0	7.1	27	0	7.1
101–200	32	0	6.4	32	0	6.4	32	0	6.4	32	0	6.4	32	0	6.4	32	0	6.4
201–300	33	0	6.3	33	0	6.3	33	0	6.3	33	0	6.3	33	0	6.3	65	1	5.0
301–400	34	0	6.1	34	0	6.1	34	0	6.1	70	1	4.6	70	1	4.6	70	1	4.6
401–500	35	0	6.1	35	0	6.1	35	0	6.1	70	1	4.7	70	1	4.7	70	1	4.7
501–600	35	0	6.1	35	0	6.1	75	1	4.4	75	1	4.4	75	1	4.4	75	1	4.4
601–800	35	0	6.2	35	0	6.2	75	1	4.4	75	1	4.4	75	1	4.4	120	2	4.2
801–1000	35	0	6.3	35	0	6.3	80	1	4.4	80	1	4.4	120	2	4.3	120	2	4.3
1001–2000	36	0	6.2	80	1	4.5	80	1	4.5	130	2	4.0	130	2	4.0	180	3	3.7
2001–3000	36	0	6.2	80	1	4.6	80	1	4.6	130	2	4.0	185	3	3.6	235	4	3.3
3001–4000	36	0	6.2	80	1	4.7	135	2	3.9	135	2	3.9	185	3	3.6	295	5	3.1
4001–5000	36	0	6.2	85	1	4.6	135	2	3.9	190	3	3.5	245	4	3.2	300	5	3.1
5001–7000	37	0	6.1	85	1	4.6	135	2	3.9	190	3	3.5	305	5	3.0	420	7	2.8
7001–10,000	37	0	6.2	85	1	4.6	135	2	3.9	245	4	3.2	310	5	3.0	430	7	2.7
10,001–20,000	85	1	4.6	135	2	3.9	195	3	3.4	250	4	3.2	435	7	2.7	635	10	2.4
20,001–50,000	85	1	4.6	135	2	3.9	255	4	3.1	380	6	2.8	575	9	2.5	990	15	2.1
50,001–100,000	85	1	4.6	135	2	3.9	255	4	3.1	445	7	2.6	790	12	2.3	1520	22	1.9

1.5%

Single Sampling Table for
Average Outgoing Quality Limit (AOQL) = 1.5%

Lot Size	Process Average 0 to 0.03%			Process Average 0.04 to 0.30%			Process Average 0.31 to 0.60%			Process Average 0.61 to 0.90%			Process Average 0.91 to 1.20%			Process Average 1.21 to 1.50%		
	n	c	p_t %	n	c	p_t %	n	c	p_t %	n	c	p_t %	n	c	p_t %	n	c	p_t %
1–15	All	0	–	All	0	–	All	0	–	All	0	–	All	0	–	All	0	–
16–50	16	0	11.6	16	0	11.6	16	0	11.6	16	0	11.6	16	0	11.6	16	0	11.6
51–100	20	0	9.8	20	0	9.8	20	0	9.8	20	0	9.8	20	0	9.8	20	0	9.8
101–200	22	0	9.5	22	0	9.5	22	0	9.5	22	0	9.5	22	0	9.5	44	1	8.2
201–300	23	0	9.2	23	0	9.2	23	0	9.2	47	1	7.9	47	1	7.9	47	1	7.9
301–400	23	0	9.3	23	0	9.3	49	1	7.8	49	1	7.8	49	1	7.8	49	1	7.8
401–500	23	0	9.4	23	0	9.4	50	1	7.7	50	1	7.7	50	1	7.7	50	1	7.7
501–600	24	0	9.0	24	0	9.0	50	1	7.7	50	1	7.7	50	1	7.7	50	1	7.7
601–800	24	0	9.1	24	0	9.1	50	1	7.8	50	1	7.8	80	2	6.4	80	2	6.4
801–1000	24	0	9.1	55	1	7.0	55	1	7.0	85	2	6.2	85	2	6.2	85	2	6.2
1001–2000	24	0	9.1	55	1	7.0	55	1	7.0	85	2	6.2	120	3	5.4	155	4	5.0
2001–3000	24	0	9.2	55	1	7.1	90	2	5.9	125	3	5.3	160	4	4.9	200	5	4.6
3001–4000	24	0	9.2	55	1	7.1	90	2	5.9	125	3	5.3	165	4	4.8	240	6	4.4
4001–5000	24	0	9.2	55	1	7.1	90	2	5.9	125	3	5.3	205	5	4.6	280	7	4.2
5001–7000	24	0	9.2	55	1	7.1	90	2	5.9	165	4	4.8	205	5	4.6	325	8	4.0
7001–10,000	24	0	9.2	55	1	7.1	130	3	5.2	165	4	4.8	250	6	4.2	375	9	3.8
10,001–20,000	55	1	7.1	90	2	5.9	130	3	5.2	210	5	4.4	340	8	3.8	515	12	3.4
20,001–50,000	55	1	7.1	90	2	5.9	170	4	4.7	295	7	4.0	480	11	3.5	860	19	3.0
50,001–100,000	55	1	7.1	130	3	5.2	210	5	4.4	340	8	3.8	625	14	3.3	1120	24	2.8

n = sample size; c = acceptance number
"All" indicates that each piece in the lot is to be inspected
p_t = lot tolerance per cent defective with a Consumer's Risk (P_C) of 0.10

Single Sampling Table for Average Outgoing Quality Limit (AOQL) = 2.0%

Lot Size	Process Average 0 to 0.04%			Process Average 0.05 to 0.40%			Process Average 0.41 to 0.80%			Process Average 0.81 to 1.20%			Process Average 1.21 to 1.60%			Process Average 1.61 to 2.00%		
	n	c	p_t %	n	c	p_t %	n	c	p_t %	n	c	p_t %	n	c	p_t %	n	c	p_t %
1–15	All	0	–	All	0	–	All	0	–	All	0	–	All	0	–	All	0	–
16–50	14	0	13.6	14	0	13.6	14	0	13.6	14	0	13.6	14	0	13.6	14	0	13.6
51–100	16	0	12.4	16	0	12.4	16	0	12.4	16	0	12.4	16	0	12.4	16	0	12.4
101–200	17	0	12.2	17	0	12.2	17	0	12.2	17	0	12.2	35	1	10.5	35	1	10.5
201–300	17	0	12.3	17	0	12.3	17	0	12.3	37	1	10.2	37	1	10.2	37	1	10.2
301–400	18	0	11.8	18	0	11.8	38	1	10.0	38	1	10.0	38	1	10.0	60	2	8.5
401–500	18	0	11.9	18	0	11.9	39	1	9.8	39	1	9.8	60	2	8.6	60	2	8.6
501–600	18	0	11.9	18	0	11.9	39	1	9.8	39	1	9.8	60	2	8.6	60	2	8.6
601–800	18	0	11.9	40	1	9.6	40	1	9.6	65	2	8.0	65	2	8.0	85	3	7.5
801–1000	18	0	12.0	40	1	9.6	40	1	9.6	65	2	8.1	65	2	8.1	90	3	7.4
1001–2000	18	0	12.0	41	1	9.4	65	2	8.2	65	2	8.2	95	3	7.0	120	4	6.5
2001–3000	18	0	12.0	41	1	9.4	65	2	8.2	95	3	7.0	120	4	6.5	180	6	5.8
3001–4000	18	0	12.0	42	1	9.3	65	2	8.2	95	3	7.0	155	5	6.0	210	7	5.5
4001–5000	18	0	12.0	42	1	9.3	70	2	7.5	125	4	6.4	155	5	6.0	245	8	5.3
5001–7000	18	0	12.0	42	1	9.3	95	3	7.0	125	4	6.4	185	6	5.6	280	9	5.1
7001–10,000	42	1	9.3	70	2	7.5	95	3	7.0	155	5	6.0	220	7	5.4	350	11	4.8
10,001–20,000	42	1	9.3	70	2	7.6	95	3	7.0	190	6	5.6	290	9	4.9	460	14	4.4
20,001–50,000	42	1	9.3	70	2	7.6	125	4	6.4	220	7	5.4	395	12	4.5	720	21	3.9
50,001–100,000	42	1	9.3	95	3	7.0	160	5	5.9	290	9	4.9	505	15	4.2	955	27	3.7

Single Sampling Table for Average Outgoing Quality Limit (AOQL) = 2.5%

2.5%

Lot Size	Process Average 0 to 0.05%			Process Average 0.06 to 0.50%			Process Average 0.51 to 1.00%			Process Average 1.01 to 1.50%			Process Average 1.51 to 2.00%			Process Average 2.01 to 2.50%		
	n	c	p_t %	n	c	p_t %	n	c	p_t %	n	c	p_t %	n	c	p_t %	n	c	p_t %
1–10	All	0	–	All	0	–	All	0	–	All	0	–	All	0	–	All	0	–
11–50	11	0	17.6	11	0	17.6	11	0	17.6	11	0	17.6	11	0	17.6	11	0	17.6
51–100	13	0	15.3	13	0	15.3	13	0	15.3	13	0	15.3	13	0	15.3	13	0	15.3
101–200	14	0	14.7	14	0	14.7	14	0	14.7	29	1	12.9	29	1	12.9	29	1	12.9
201–300	14	0	14.9	14	0	14.9	30	1	12.7	30	1	12.7	30	1	12.7	30	1	12.7
301–400	14	0	15.0	14	0	15.0	31	1	12.3	31	1	12.3	31	1	12.3	48	2	10.7
401–500	14	0	15.0	14	0	15.0	32	1	12.0	32	1	12.0	49	2	10.6	49	2	10.6
501–600	14	0	15.1	32	1	12.0	32	1	12.0	50	2	10.4	50	2	10.4	70	3	9.3
601–800	14	0	15.1	32	1	12.0	32	1	12.0	50	2	10.5	50	2	10.5	70	3	9.4
801–1000	15	0	14.2	33	1	11.7	33	1	11.7	50	2	10.6	70	3	9.4	90	4	8.5
1001–2000	15	0	14.2	33	1	11.7	55	2	9.3	75	3	8.8	95	4	8.0	120	5	7.6
2001–3000	15	0	14.2	33	1	11.8	55	2	9.4	75	3	8.8	120	5	7.6	145	6	7.2
3001–4000	15	0	14.3	33	1	11.8	55	2	9.5	100	4	7.9	125	5	7.4	195	8	6.6
4001–5000	15	0	14.3	33	1	11.8	75	3	8.9	100	4	7.9	150	6	7.0	225	9	6.3
5001–7000	33	1	11.8	55	2	9.7	75	3	8.9	125	5	7.4	175	7	6.7	250	10	6.1
7001–10,000	34	1	11.4	55	2	9.7	75	3	8.9	125	5	7.4	200	8	6.4	310	12	5.8
10,001–20,000	34	1	11.4	55	2	9.7	100	4	8.0	150	6	7.0	260	10	6.0	425	16	5.3
20,001–50,000	34	1	11.4	55	2	9.7	100	4	8.0	180	7	6.7	345	13	5.5	640	23	4.8
50,001–100,000	34	1	11.4	80	3	8.4	125	5	7.4	235	9	6.1	435	16	5.2	800	28	4.5

n = sample size; c = acceptance number
"All" indicates that each piece in the lot is to be inspected
p_t = lot tolerance per cent defective with a Consumer's Risk (P_C) of 0.10

SINGLE SAMPLING

3.0%
AOQL

Single Sampling Table for
Average Outgoing Quality Limit (AOQL) = 3.0%

Lot Size	Process Average 0 to 0.06%			Process Average 0.07 to 0.60%			Process Average 0.61 to 1.20%			Process Average 1.21 to 1.80%			Process Average 1.81 to 2.40%			Process Average 2.41 to 3.00%		
	n	c	p_t %	n	c	p_t %	n	c	p_t %	n	c	p_t %	n	c	p_t %	n	c	p_t %
1–10	All	0	–	All	0	–	All	0	–	All	0	–	All	0	–	All	0	–
11–50	10	0	19.0	10	0	19.0	10	0	19.0	10	0	19.0	10	0	19.0	10	0	19.0
51–100	11	0	18.0	11	0	18.0	11	0	18.0	11	0	18.0	11	0	18.0	22	1	16.4
101–200	12	0	17.0	12	0	17.0	12	0	17.0	25	1	15.1	25	1	15.1	25	1	15.1
201–300	12	0	17.0	12	0	17.0	26	1	14.6	26	1	14.6	26	1	14.6	40	2	12.8
301–400	12	0	17.1	12	0	17.1	26	1	14.7	26	1	14.7	41	2	12.7	41	2	12.7
401–500	12	0	17.2	27	1	14.1	27	1	14.1	42	2	12.4	42	2	12.4	42	2	12.4
501–600	12	0	17.3	27	1	14.2	27	1	14.2	42	2	12.4	42	2	12.4	60	3	10.8
601–800	12	0	17.3	27	1	14.2	27	1	14.2	43	2	12.1	60	3	10.9	60	3	10.9
801–1000	12	0	17.4	27	1	14.2	44	2	11.8	44	2	11.8	60	3	11.0	80	4	9.8
1001–2000	12	0	17.5	28	1	13.8	45	2	11.7	65	3	10.2	80	4	9.8	100	5	9.1
2001–3000	12	0	17.5	28	1	13.8	45	2	11.7	65	3	10.2	100	5	9.1	140	7	8.2
3001–4000	12	0	17.5	28	1	13.8	65	3	10.3	85	4	9.5	125	6	8.4	165	8	7.8
4001–5000	28	1	13.8	28	1	13.8	65	3	10.3	85	4	9.5	125	6	8.4	210	10	7.4
5001–7000	28	1	13.8	45	2	11.8	65	3	10.3	105	5	8.8	145	7	8.1	235	11	7.1
7001–10,000	28	1	13.9	46	2	11.6	65	3	10.3	105	5	8.8	170	8	7.6	280	13	6.8
10,001–20,000	28	1	13.9	46	2	11.7	85	4	9.5	125	6	8.4	215	10	7.2	380	17	6.2
20,001–50,000	28	1	13.9	65	3	10.3	105	5	8.8	170	8	7.6	310	14	6.5	560	24	5.7
50,001–100,000	28	1	13.9	65	3	10.3	125	6	8.4	215	10	7.2	385	17	6.2	690	29	5.4

4.0%

Single Sampling Table for
Average Outgoing Quality Limit (AOQL) = 4.0%

Lot Size	Process Average 0 to 0.08%			Process Average 0.09 to 0.80%			Process Average 0.81 to 1.60%			Process Average 1.61 to 2.40%			Process Average 2.41 to 3.20%			Process Average 3.21 to 4.00%		
	n	c	p_t %	n	c	p_t %	n	c	p_t %	n	c	p_t %	n	c	p_t %	n	c	p_t %
1–10	All	0	–	All	0	–	All	0	–	All	0	–	All	0	–	All	0	–
11–50	8	0	23.0	8	0	23.0	8	0	23.0	8	0	23.0	8	0	23.0	8	0	23.0
51–100	8	0	24.0	8	0	24.0	8	0	24.0	8	0	24.0	17	1	21.5	17	1	21.5
101–200	9	0	22.0	9	0	22.0	19	1	20.0	19	1	20.0	19	1	20.0	19	1	20.0
201–300	9	0	22.5	9	0	22.5	20	1	19.0	20	1	19.0	31	2	16.8	31	2	16.8
301–400	9	0	22.5	20	1	19.1	20	1	19.1	32	2	16.2	32	2	16.2	43	3	15.2
401–500	9	0	22.5	20	1	19.1	20	1	19.1	32	2	16.3	32	2	16.3	44	3	14.9
501–600	9	0	22.5	20	1	19.2	20	1	19.2	32	2	16.3	45	3	14.6	60	4	12.9
601–800	9	0	22.5	20	1	19.2	33	2	15.9	33	2	15.9	46	3	14.3	60	4	13.0
801–1000	9	0	22.5	21	1	18.3	33	2	16.0	46	3	14.3	60	4	13.0	75	5	12.2
1001–2000	9	0	22.5	21	1	18.4	34	2	15.6	47	3	14.1	75	5	12.2	105	7	11.0
2001–3000	9	0	22.5	21	1	18.4	34	2	15.6	60	4	13.2	90	6	11.3	125	8	10.4
3001–4000	21	1	18.4	21	1	18.4	48	3	13.8	65	4	12.2	110	7	10.7	155	10	9.8
4001–5000	21	1	18.5	34	2	15.7	48	3	13.9	80	5	11.6	110	7	10.8	175	11	9.5
5001–7000	21	1	18.5	34	2	15.7	48	3	13.9	80	5	11.6	125	8	10.4	210	13	9.0
7001–10,000	21	1	18.5	34	2	15.7	65	4	12.3	95	6	11.1	145	9	9.8	245	15	8.6
10,001–20,000	21	1	18.5	34	2	15.7	65	4	12.3	110	7	10.8	195	12	9.0	340	20	7.9
20,001–50,000	21	1	18.5	49	3	13.6	80	5	11.6	145	9	9.8	250	15	8.5	460	26	7.4
50,001–100,000	21	1	18.5	49	3	13.6	95	6	11.1	165	10	9.6	310	18	8.0	540	30	7.1

n = sample size; c = acceptance number
"All" indicates that each piece in the lot is to be inspected
p_t = lot tolerance per cent defective with a Consumer's Risk (P_C) of 0.10

Single Sampling Table for
Average Outgoing Quality Limit (AOQL) = 5.0%

5.0% AOQL

Lot Size	Process Average 0 to 0.10%			Process Average 0.11 to 1.00%			Process Average 1.01 to 2.00%			Process Average 2.01 to 3.00%			Process Average 3.01 to 4.00%			Process Average 4.01 to 5.00%		
	n	c	p_t %	n	c	p_t %	n	c	p_t %	n	c	p_t %	n	c	p_t %	n	c	p_t %
1–5	All	0	–	All	0	–	All	0	–	All	0	–	All	0	–	All	0	–
6–50	6	0	30.5	6	0	30.5	6	0	30.5	6	0	30.5	6	0	30.5	6	0	30.5
51–100	7	0	27.0	7	0	27.0	7	0	27.0	14	1	26.5	14	1	26.5	14	1	26.5
101–200	7	0	27.5	7	0	27.5	16	1	24.0	16	1	24.0	16	1	24.0	24	2	21.5
201–300	7	0	27.5	16	1	24.0	16	1	24.0	16	1	24.0	25	2	21.0	25	2	21.0
301–400	7	0	27.5	16	1	24.0	16	1	24.0	26	2	20.0	26	2	20.0	35	3	18.8
401–500	7	0	27.5	16	1	24.0	16	1	24.0	26	2	20.0	36	3	18.3	46	4	17.0
501–600	7	0	28.0	16	1	24.0	26	2	20.0	26	2	20.0	37	3	17.9	47	4	16.6
601–800	7	0	28.0	16	1	24.0	27	2	19.4	37	3	17.9	48	4	16.3	60	5	15.2
801–1000	7	0	28.0	17	1	22.5	27	2	19.5	37	3	17.9	48	4	16.3	70	6	14.3
1001–2000	7	0	28.0	17	1	23.0	27	2	19.6	38	3	17.6	60	5	15.3	85	7	13.7
2001–3000	7	0	28.0	17	1	23.0	38	3	17.6	50	4	15.8	75	6	13.9	125	10	12.3
3001–4000	17	1	23.0	27	2	19.6	39	3	17.0	60	5	15.4	85	7	13.8	140	11	11.8
4001–5000	17	1	23.0	27	2	19.6	39	3	17.0	65	5	14.2	100	8	12.9	155	12	11.6
5001–7000	17	1	23.0	27	2	19.7	39	3	17.1	75	6	13.9	115	9	12.3	185	14	11.0
7001–10,000	17	1	23.0	27	2	19.7	50	4	15.9	75	6	14.0	130	10	12.0	225	17	10.4
10,001–20,000	17	1	23.0	27	2	19.7	50	4	15.9	90	7	13.1	170	13	11.0	305	22	9.6
20,001–50,000	17	1	23.0	39	3	17.1	65	5	14.3	115	9	12.3	215	16	10.4	400	28	9.0
50,001–100,000	17	1	23.0	39	3	17.1	75	6	14.0	145	11	11.6	275	20	9.8	450	31	8.8

Single Sampling Table for
Average Outgoing Quality Limit (AOQL) = 7.0%

7.0%

Lot Size	Process Average 0 to 0.14%			Process Average 0.15 to 1.40%			Process Average 1.41 to 2.80%			Process Average 2.81 to 4.20%			Process Average 4.21 to 5.60%			Process Average 5.61 to 7.00%		
	n	c	p_t %	n	c	p_t %	n	c	p_t %	n	c	p_t %	n	c	p_t %	n	c	p_t %
1–5	All	0	–	All	0	–	All	0	–	All	0	–	All	0	–	All	0	–
6–50	5	0	35.5	5	0	35.5	5	0	35.5	5	0	35.5	5	0	35.5	5	0	35.5
51–100	5	0	36.0	5	0	36.0	5	0	36.0	11	1	28.5	11	1	28.5	11	1	28.5
101–200	5	0	36.5	5	0	36.5	11	1	30.5	11	1	30.5	18	2	26.5	18	2	26.5
201–300	5	0	36.5	12	1	28.5	12	1	28.5	18	2	26.5	18	2	26.5	25	3	26.0
301–400	5	0	37.0	12	1	28.5	12	1	28.5	19	2	25.5	26	3	25.0	33	4	23.5
401–500	5	0	37.0	12	1	28.5	19	2	25.5	19	2	25.5	26	3	25.0	34	4	23.0
501–600	5	0	37.0	12	1	28.5	19	2	25.5	27	3	24.5	34	4	23.0	42	5	21.5
601–800	5	0	37.0	12	1	29.0	19	2	25.5	27	3	24.5	35	4	22.5	50	6	20.5
801–1000	5	0	37.0	12	1	29.0	19	2	25.5	27	3	24.5	43	5	21.5	60	7	19.3
1001–2000	5	0	37.0	12	1	29.0	27	3	24.5	36	4	22.0	50	6	21.0	70	8	17.7
2001–3000	12	1	29.0	19	2	25.5	28	3	23.5	45	5	20.5	60	7	19.6	100	11	16.5
3001–4000	12	1	29.0	20	2	24.5	28	3	24.0	45	5	20.5	70	8	18.1	120	13	15.8
4001–5000	12	1	29.0	20	2	24.5	36	4	22.0	55	6	19.0	80	9	17.3	140	15	15.1
5001–7000	12	1	29.0	20	2	24.5	36	4	22.0	55	6	19.1	90	10	16.8	160	17	14.6
7001–10,000	12	1	29.0	20	2	24.5	36	4	22.0	65	7	18.4	110	12	15.9	195	20	13.9
10,001–20,000	12	1	29.0	28	3	24.0	45	5	20.5	75	8	17.8	135	14	15.2	240	24	13.2
20,001–50,000	12	1	29.0	28	3	24.0	55	6	19.2	95	10	16.6	175	18	14.1	310	30	12.4
50,001–100,000	12	1	29.0	28	3	24.0	55	6	19.2	115	12	15.8	210	21	13.4	355	34	12.1

n = sample size; c = acceptance number
"All" indicates that each piece in the lot is to be inspected
p_t = lot tolerance per cent defective with a Consumer's Risk (P_C) of 0.10

SINGLE
SAMPLING

10.0%

A O Q L

— 204 —

Appendix 6

Single Sampling Table for
Average Outgoing Quality Limit (AOQL) = 10.0%

Lot Size	Process Average 0 to 0.20%			Process Average 0.21 to 2.00%			Process Average 2.01 to 4.00%			Process Average 4.01 to 6.00%			Process Average 6.01 to 8.00%			Process Average 8.01 to 10.00%		
	n	c	p_t %	n	c	p_t %	n	c	p_t %	n	c	p_t %	n	c	p_t %	n	c	p_t %
1–3	All	0	–	All	0	–	All	0	–	All	0	–	All	0	–	All	0	–
4–50	3	0	52.5	3	0	52.5	3	0	52.5	3	0	52.5	3	0	52.5	7	1	43.5
51–100	4	0	43.0	4	0	43.0	8	1	40.0	8	1	40.0	8	1	40.0	12	2	37.5
101–200	4	0	43.5	8	1	40.0	8	1	40.0	13	2	35.5	13	2	35.5	18	3	33.0
201–300	4	0	43.5	8	1	40.5	8	1	40.5	13	2	35.5	18	3	33.0	23	4	32.0
301–400	4	0	43.5	8	1	40.5	13	2	35.5	13	2	35.5	24	4	30.0	29	5	30.0
401–500	4	0	43.5	8	1	40.5	13	2	36.0	19	3	31.5	24	4	30.0	30	5	29.5
501–600	4	0	43.5	8	1	40.5	13	2	36.0	19	3	31.5	24	4	30.5	36	6	28.5
601–800	4	0	43.5	8	1	40.5	13	2	36.0	19	3	31.5	31	5	29.5	42	7	27.5
801–1000	4	0	44.0	8	1	40.5	14	2	33.5	25	4	30.0	37	6	28.0	49	8	26.5
1001–2000	8	1	40.5	14	2	33.5	19	3	32.0	31	5	30.0	44	7	26.5	65	10	23.5
2001–3000	8	1	40.5	14	2	33.5	19	3	32.0	31	5	30.0	50	8	26.0	85	13	22.5
3001–4000	8	1	40.5	14	2	33.5	25	4	30.0	38	6	27.5	65	10	24.0	100	15	21.5
4001–5000	8	1	40.5	14	2	33.5	25	4	30.0	38	6	27.5	65	10	24.0	120	18	20.5
5001–7000	8	1	40.5	14	2	33.5	25	4	30.0	44	7	27.0	80	12	22.5	135	20	19.8
7001–10,000	8	1	40.5	14	2	33.5	32	5	29.0	50	8	26.0	85	13	22.5	160	23	19.2
10,001–20,000	8	1	40.5	19	3	32.0	32	5	29.0	60	9	24.5	110	16	21.0	190	27	18.3
20,001–50,000	8	1	40.5	19	3	32.0	38	6	27.5	70	11	23.0	130	19	19.7	225	31	17.5
50,001–100,000	14	2	33.5	19	3	32.0	44	7	27.0	80	12	22.5	155	22	19.0	260	35	16.9

n = sample size; c = acceptance number
"All" indicates that each piece in the lot is to be inspected
p_t = lot tolerance per cent defective with a Consumer's Risk (P_C) of 0.10

Appendix 7

Double Sampling Tables for Stated Values of
Average Outgoing Quality Limit (AOQL)

DOUBLE SAMPLING
0.1%
AOQL

Double Sampling Table for Average Outgoing Quality Limit (AOQL) = 0.1%

Lot Size	Process Average 0 to 0.002% Trial 1 n_1	c_1	Trial 2 n_2	n_1+n_2	c_2	p_t %	Process Average 0.003 to 0.020% Trial 1 n_1	c_1	Trial 2 n_2	n_1+n_2	c_2	p_t %	Process Average 0.021 to 0.040% Trial 1 n_1	c_1	Trial 2 n_2	n_1+n_2	c_2	p_t %
1–75	All	0	–	–	–	–	All	0	–	–	–	–	All	0	–	–	–	–
76–95	75	0	–	–	–	1.5	75	0	–	–	–	1.5	75	0	–	–	–	1.5
96–130	95	0	–	–	–	1.4	95	0	–	–	–	1.4	95	0	–	–	–	1.4
131–200	130	0	–	–	–	1.2	130	0	–	–	–	1.2	130	0	–	–	–	1.2
201–300	165	0	–	–	–	1.1	165	0	–	–	–	1.1	165	0	–	–	–	1.1
301–350	190	0	–	–	–	0.96	190	0	–	–	–	0.96	190	0	–	–	–	0.96
351–400	225	0	95	320	1	0.86	225	0	95	320	1	0.86	225	0	95	320	1	0.86
401–500	250	0	120	370	1	0.80	250	0	120	370	1	0.80	250	0	120	370	1	0.80
501–600	275	0	130	405	1	0.77	275	0	130	405	1	0.77	275	0	130	405	1	0.77
601–800	310	0	155	465	1	0.71	310	0	155	465	1	0.71	310	0	155	465	1	0.71
801–1000	350	0	185	535	1	0.66	350	0	185	535	1	0.66	350	0	185	535	1	0.66
1001–2000	430	0	240	670	1	0.58	430	0	240	670	1	0.58	430	0	240	670	1	0.58
2001–3000	465	0	265	730	1	0.56	465	0	265	730	1	0.56	465	0	265	730	1	0.56
3001–4000	495	0	285	780	1	0.54	495	0	285	780	1	0.54	540	0	570	1110	2	0.49
4001–5000	505	0	295	800	1	0.53	505	0	295	800	1	0.53	555	0	615	1170	2	0.48
5001–7000	520	0	320	840	1	0.52	520	0	320	840	1	0.52	590	0	660	1250	2	0.46
7001–10,000	540	0	335	875	1	0.51	625	0	715	1340	2	0.44	625	0	715	1340	2	0.44
10,001–20,000	555	0	345	900	1	0.50	650	0	750	1400	2	0.43	720	0	1150	1870	3	0.38
20,001–50,000	660	0	760	1420	2	0.42	660	0	760	1420	2	0.42	770	0	1700	2470	4	0.36
50,001–100,000	670	0	770	1440	2	0.42	740	0	1230	1970	3	0.38	805	0	1725	2530	4	0.35

Lot Size	Process Average 0.041 to 0.060% Trial 1 n_1	c_1	Trial 2 n_2	n_1+n_2	c_2	p_t %	Process Average 0.061 to 0.080% Trial 1 n_1	c_1	Trial 2 n_2	n_1+n_2	c_2	p_t %	Process Average 0.081 to 0.100% Trial 1 n_1	c_1	Trial 2 n_2	n_1+n_2	c_2	p_t %
1–75	All	0	–	–	–	–	All	0	–	–	–	–	All	0	–	–	–	–
76–95	75	0	–	–	–	1.5	75	0	–	–	–	1.5	75	0	–	–	–	1.5
96–130	95	0	–	–	–	1.4	95	0	–	–	–	1.4	95	0	–	–	–	1.4
131–200	130	0	–	–	–	1.2	130	0	–	–	–	1.2	130	0	–	–	–	1.2
201–300	165	0	–	–	–	1.1	165	0	–	–	–	1.1	165	0	–	–	–	1.1
301–350	190	0	–	–	–	0.96	190	0	–	–	–	0.96	190	0	–	–	–	0.96
351–400	225	0	95	320	1	0.86	225	0	95	320	1	0.86	225	0	95	320	1	0.86
401–500	250	0	120	370	1	0.80	250	0	120	370	1	0.80	250	0	120	370	1	0.80
501–600	275	0	130	405	1	0.77	275	0	130	405	1	0.77	275	0	130	405	1	0.77
601–800	310	0	155	465	1	0.71	310	0	155	465	1	0.71	310	0	155	465	1	0.71
801–1000	350	0	185	535	1	0.66	350	0	185	535	1	0.66	350	0	185	535	1	0.66
1001–2000	430	0	240	670	1	0.58	430	0	240	670	1	0.58	475	0	450	925	2	0.54
2001–3000	520	0	530	1050	2	0.50	520	0	530	1050	2	0.50	520	0	530	1050	2	0.50
3001–4000	540	0	570	1110	2	0.49	540	0	570	1110	2	0.49	585	0	885	1470	3	0.45
4001–5000	555	0	615	1170	2	0.48	605	0	935	1540	3	0.44	605	0	935	1540	3	0.44
5001–7000	655	0	1045	1700	3	0.41	655	0	1045	1700	3	0.41	680	0	1360	2040	4	0.40
7001–10,000	700	0	1120	1820	3	0.39	700	0	1120	1820	3	0.39	1250	1	1780	3030	6	0.35
10,001–20,000	740	0	1530	2270	4	0.37	1350	1	2020	3370	6	0.33	1400	1	2420	3820	7	0.32
20,001–50,000	1400	1	2170	3570	6	0.32	1450	1	3030	4480	8	0.31	2230	2	4650	6880	12	0.27
50,001–100,000	1460	1	3060	4520	8	0.31	2330	2	4910	7240	12	0.26	3690	4	7580	11270	19	0.24

Trial 1: n_1 = first sample size; c_1 = acceptance number for first sample
"All" indicates that each piece in the lot is to be inspected
Trial 2: n_2 = second sample size; c_2 = acceptance number for first and second samples combined
p_t = lot tolerance per cent defective with a Consumer's Risk (P_C) of 0.10

Double Sampling Table for
Average Outgoing Quality Limit (AOQL) = 0.25%

Lot Size	Process Average 0 to 0.005% Trial 1 n_1	c_1	Trial 2 n_2	n_1+n_2	c_2	p_t %	Process Average 0.006 to 0.050% Trial 1 n_1	c_1	Trial 2 n_2	n_1+n_2	c_2	p_t %	Process Average 0.051 to 0.100% Trial 1 n_1	c_1	Trial 2 n_2	n_1+n_2	c_2	p_t %
1–60	All	0	–	–	–	–	All	0	–	–	–	–	All	0	–	–	–	–
61–100	60	0	–	–	–	2.5	60	0	–	–	–	2.5	60	0	–	–	–	2.5
101–200	85	0	–	–	–	2.1	85	0	–	–	–	2.1	85	0	–	–	–	2.1
201–300	120	0	65	185	1	1.8	120	0	65	185	1	1.8	120	0	65	185	1	1.8
301–400	135	0	70	205	1	1.7	135	0	70	205	1	1.7	135	0	70	205	1	1.7
401–500	145	0	80	225	1	1.6	145	0	80	225	1	1.6	145	0	80	225	1	1.6
501–600	160	0	90	250	1	1.5	160	0	90	250	1	1.5	160	0	90	250	1	1.5
601–800	165	0	95	260	1	1.5	165	0	95	260	1	1.5	165	0	95	260	1	1.5
801–1000	180	0	105	285	1	1.4	180	0	105	285	1	1.4	180	0	105	285	1	1.4
1001–2000	205	0	120	325	1	1.3	205	0	120	325	1	1.3	220	0	245	465	2	1.2
2001–3000	210	0	125	335	1	1.3	210	0	125	335	1	1.3	235	0	275	510	2	1.1
3001–4000	210	0	130	340	1	1.3	210	0	130	340	1	1.3	240	0	280	520	2	1.1
4001–5000	215	0	130	345	1	1.3	245	0	280	525	2	1.1	275	0	445	720	3	1.0
5001–7000	215	0	135	350	1	1.3	250	0	285	535	2	1.1	290	0	475	765	3	0.95
7001–10,000	255	0	290	545	2	1.1	255	0	290	545	2	1.1	295	0	490	785	3	0.94
10,001–20,000	260	0	295	555	2	1.1	260	0	295	555	2	1.1	330	0	730	1060	4	0.83
20,001–50,000	265	0	300	565	2	1.0	305	0	505	810	3	0.91	335	0	735	1070	4	0.83
50,001–100,000	270	0	305	575	2	1.0	310	0	510	820	3	0.91	350	0	930	1280	5	0.80

Lot Size	Process Average 0.101 to 0.150% Trial 1 n_1	c_1	Trial 2 n_2	n_1+n_2	c_2	p_t %	Process Average 0.151 to 0.200% Trial 1 n_1	c_1	Trial 2 n_2	n_1+n_2	c_2	p_t %	Process Average 0.201 to 0.250% Trial 1 n_1	c_1	Trial 2 n_2	n_1+n_2	c_2	p_t %
1–60	All	0	–	–	–	–	All	0	–	–	–	–	All	0	–	–	–	–
61–100	60	0	–	–	–	2.5	60	0	–	–	–	2.5	60	0	–	–	–	2.5
101–200	85	0	–	–	–	2.1	85	0	–	–	–	2.1	85	0	–	–	–	2.1
201–300	120	0	65	185	1	1.8	120	0	65	185	1	1.8	120	0	65	185	1	1.8
301–400	135	0	70	205	1	1.7	135	0	70	205	1	1.7	135	0	70	205	1	1.7
401–500	145	0	80	225	1	1.6	145	0	80	225	1	1.6	145	0	80	225	1	1.6
501–600	160	0	90	250	1	1.5	160	0	90	250	1	1.5	175	0	160	335	2	1.4
601–800	165	0	95	260	1	1.5	195	0	185	380	2	1.3	195	0	185	380	2	1.3
801–1000	200	0	195	395	2	1.3	200	0	195	395	2	1.3	200	0	195	395	2	1.3
1001–2000	220	0	245	465	2	1.2	220	0	245	465	2	1.2	240	0	375	615	3	1.1
2001–3000	260	0	435	695	3	1.0	260	0	435	695	3	1.0	280	0	570	850	4	0.96
3001–4000	270	0	440	710	3	1.0	290	0	600	890	4	0.94	290	0	600	890	4	0.94
4001–5000	275	0	445	720	3	1.0	300	0	615	915	4	0.92	520	1	750	1270	6	0.85
5001–7000	315	0	660	975	4	0.88	545	1	795	1340	6	0.82	555	1	965	1520	7	0.80
7001–10,000	325	0	705	1030	4	0.84	585	1	1045	1630	7	0.76	620	1	1420	2040	9	0.72
10,001–20,000	590	1	910	1500	6	0.76	660	1	1530	2190	9	0.68	1220	3	2000	3220	13	0.61
20,001–50,000	645	1	1355	2000	8	0.70	990	2	2340	3330	13	0.61	1850	5	3500	5350	21	0.55
50,001–100,000	665	1	1615	2280	9	0.68	1340	3	2910	4250	16	0.56	2930	8	6090	9020	33	0.48

Trial 1: n_1 = first sample size; c_1 = acceptance number for first sample
"All" indicates that each piece in the lot is to be inspected
Trial 2: n_2 = second sample size; c_2 = acceptance number for first and second samples combined
p_t = lot tolerance per cent defective with a Consumer's Risk (P_C) of 0.10

DOUBLE
SAMPLING

0.5%
AOQL

Double Sampling Table for
Average Outgoing Quality Limit (AOQL) = 0.5%

Lot Size	Process Average 0 to 0.010%						Process Average 0.011 to 0.10%						Process Average 0.11 to 0.20%					
	Trial 1		Trial 2			p_t %	Trial 1		Trial 2			p_t %	Trial 1		Trial 2			p_t %
	n_1	c_1	n_2	n_1+n_2	c_2		n_1	c_1	n_2	n_1+n_2	c_2		n_1	c_1	n_2	n_1+n_2	c_2	
1–30	All	0	–	–	–	–	All	0	–	–	–	–	All	0	–	–	–	–
31–50	30	0	–	–	–	5.0	30	0	–	–	–	5.0	30	0	–	–	–	5.0
51–75	40	0	–	–	–	4.6	40	0	–	–	–	4.6	40	0	–	–	–	4.6
76–100	47	0	23	70	1	4.4	47	0	23	70	1	4.4	47	0	23	70	1	4.4
101–150	60	0	30	90	1	3.8	60	0	30	90	1	3.8	60	0	30	90	1	3.8
151–200	70	0	35	105	1	3.3	70	0	35	105	1	3.3	70	0	35	105	1	3.3
201–300	80	0	45	125	1	3.0	80	0	45	125	1	3.0	80	0	45	125	1	3.0
301–400	85	0	50	135	1	2.9	85	0	50	135	1	2.9	85	0	50	135	1	2.9
401–500	90	0	55	145	1	2.8	90	0	55	145	1	2.8	90	0	55	145	1	2.8
501–600	95	0	55	150	1	2.8	95	0	55	150	1	2.8	95	0	55	150	1	2.8
601–800	100	0	55	155	1	2.7	100	0	55	155	1	2.7	110	0	115	225	2	2.4
801–1000	100	0	60	160	1	2.7	100	0	60	160	1	2.7	115	0	125	240	2	2.3
1001–2000	105	0	60	165	1	2.6	125	0	135	260	2	2.2	125	0	135	260	2	2.2
2001–3000	110	0	60	170	1	2.6	130	0	145	275	2	2.1	145	0	235	380	3	1.9
3001–4000	110	0	65	175	1	2.5	130	0	155	285	2	2.1	145	0	240	385	3	1.9
4001–5000	135	0	150	285	2	2.1	135	0	150	285	2	2.1	150	0	240	390	3	1.9
5001–7000	135	0	155	290	2	2.1	135	0	155	290	2	2.1	155	0	245	400	3	1.8
7001–10,000	135	0	160	295	2	2.1	135	0	160	295	2	2.1	175	0	375	550	4	1.6
10,001–20,000	140	0	160	300	2	2.0	155	0	250	405	3	1.8	185	0	500	685	5	1.5
20,001–50,000	140	0	165	305	2	2.0	155	0	255	410	3	1.8	185	0	505	690	5	1.5
50,001–100,000	140	0	170	310	2	2.0	155	0	260	415	3	1.8	325	1	495	820	6	1.4

Lot Size	Process Average 0.21 to 0.30%						Process Average 0.31 to 0.40%						Process Average 0.41 to 0.50%					
	Trial 1		Trial 2			p_t %	Trial 1		Trial 2			p_t %	Trial 1		Trial 2			p_t %
	n_1	c_1	n_2	n_1+n_2	c_2		n_1	c_1	n_2	n_1+n_2	c_2		n_1	c_1	n_2	n_1+n_2	c_2	
1–30	All	0	–	–	–	–	All	0	–	–	–	–	All	0	–	–	–	–
31–50	30	0	–	–	–	5.0	30	0	–	–	–	5.0	30	0	–	–	–	5.0
51–75	40	0	–	–	–	4.6	40	0	–	–	–	4.6	40	0	–	–	–	4.6
76–100	47	0	23	70	1	4.4	47	0	23	70	1	4.4	47	0	23	70	1	4.4
101–150	60	0	30	90	1	3.8	60	0	30	90	1	3.8	60	0	30	90	1	3.8
151–200	70	0	35	105	1	3.3	70	0	35	105	1	3.3	70	0	35	105	1	3.3
201–300	80	0	45	125	1	3.0	80	0	45	125	1	3.0	80	0	45	125	1	3.0
301–400	85	0	50	135	1	2.9	95	0	90	185	2	2.7	95	0	90	185	2	2.7
401–500	100	0	100	200	2	2.6	100	0	100	200	2	2.6	100	0	100	200	2	2.6
501–600	105	0	105	210	2	2.5	105	0	105	210	2	2.5	105	0	105	210	2	2.5
601–800	110	0	115	225	2	2.4	110	0	115	225	2	2.4	120	0	180	300	3	2.2
801–1000	115	0	125	240	2	2.3	125	0	185	310	3	2.2	125	0	185	310	3	2.2
1001–2000	135	0	220	355	3	2.0	135	0	220	355	3	2.0	145	0	295	440	4	1.9
2001–3000	145	0	235	380	3	1.9	150	0	320	470	4	1.8	275	1	475	750	7	1.6
3001–4000	155	0	325	480	4	1.8	280	1	415	695	6	1.6	295	1	600	895	8	1.5
4001–5000	165	0	345	510	4	1.7	300	1	525	825	7	1.5	430	2	700	1130	10	1.4
5001–7000	175	0	455	630	5	1.6	310	1	670	980	8	1.4	460	2	860	1320	11	1.3
7001–10,000	300	1	460	760	6	1.5	465	2	785	1250	10	1.3	620	3	1120	1740	14	1.2
10,001–20,000	320	1	680	1000	8	1.4	495	2	1175	1670	13	1.2	740	4	1420	2160	18	1.2
20,001–50,000	350	1	930	1280	10	1.3	680	3	1490	2170	16	1.1	925	5	2085	3010	24	1.1
50,001–100,000	505	2	1075	1580	12	1.2	680	3	1810	2490	19	1.1	1550	9	3410	4960	38	0.99

Trial 1: n_1 = first sample size; c_1 = acceptance number for first sample
"All" indicates that each piece in the lot is to be inspected
Trial 2: n_2 = second sample size; c_2 = acceptance number for first and second samples combined
p_t = lot tolerance per cent defective with a Consumer's Risk (P_C) of 0.10

Double Sampling Table for Average Outgoing Quality Limit (AOQL) = 0.75%

Lot Size	Process Average 0 to 0.015% Trial 1 n_1	c_1	Trial 2 n_2	n_1+n_2	c_2	p_t %	Process Average 0.016 to 0.15% Trial 1 n_1	c_1	Trial 2 n_2	n_1+n_2	c_2	p_t %	Process Average 0.16 to 0.30% Trial 1 n_1	c_1	Trial 2 n_2	n_1+n_2	c_2	p_t %
1–25	All	0	–	–	–	–	All	0	–	–	–	–	All	0	–	–	–	–
26–50	25	0	–	–	–	6.4	25	0	–	–	–	6.4	25	0	–	–	–	6.4
51–75	35	0	15	50	1	6.0	35	0	15	50	1	6.0	35	0	15	50	1	6.0
76–100	39	0	21	60	1	5.6	39	0	21	60	1	5.6	39	0	21	60	1	5.6
101–200	55	0	30	85	1	4.5	55	0	30	85	1	4.5	55	0	30	85	1	4.5
201–300	60	0	30	90	1	4.3	60	0	30	90	1	4.3	60	0	30	90	1	4.3
301–400	60	0	35	95	1	4.2	60	0	35	95	1	4.2	60	0	35	95	1	4.2
401–500	65	0	35	100	1	4.1	65	0	35	100	1	4.1	70	0	75	145	2	3.7
501–600	65	0	40	105	1	4.0	65	0	40	105	1	4.0	75	0	80	155	2	3.6
601–800	70	0	40	110	1	3.9	70	0	40	110	1	3.9	75	0	85	160	2	3.5
801–1000	70	0	40	110	1	3.9	70	0	40	110	1	3.9	80	0	90	170	2	3.4
1001–2000	75	0	40	115	1	3.8	85	0	90	175	2	3.3	95	0	155	250	3	2.9
2001–3000	75	0	40	115	1	3.8	85	0	100	185	2	3.2	100	0	160	260	3	2.8
3001–4000	90	0	105	195	2	3.1	90	0	105	195	2	3.1	100	0	160	260	3	2.8
4001–5000	90	0	105	195	2	3.1	90	0	105	195	2	3.1	105	0	235	340	4	2.6
5001–7000	90	0	105	195	2	3.1	90	0	105	195	2	3.1	110	0	245	355	4	2.5
7001–10,000	90	0	110	200	2	3.0	100	0	165	265	3	2.8	110	0	245	355	4	2.5
10,001–20,000	90	0	110	200	2	3.0	100	0	165	265	3	2.8	120	0	325	445	5	2.3
20,001–50,000	90	0	110	200	2	3.0	105	0	165	270	3	2.7	130	0	335	465	5	2.2
50,001–100,000	95	0	110	205	2	3.0	110	0	250	360	4	2.5	215	1	395	610	7	2.1

Lot Size	Process Average 0.31 to 0.45% Trial 1 n_1	c_1	Trial 2 n_2	n_1+n_2	c_2	p_t %	Process Average 0.46 to 0.60% Trial 1 n_1	c_1	Trial 2 n_2	n_1+n_2	c_2	p_t %	Process Average 0.61 to 0.75% Trial 1 n_1	c_1	Trial 2 n_2	n_1+n_2	c_2	p_t %
1–25	All	0	–	–	–	–	All	0	–	–	–	–	All	0	–	–	–	–
26–50	25	0	–	–	–	6.4	25	0	–	–	–	6.4	25	0	–	–	–	6.4
51–75	35	0	15	50	1	6.0	35	0	15	50	1	6.0	35	0	15	50	1	6.0
76–100	39	0	21	60	1	5.6	39	0	21	60	1	5.6	39	0	21	60	1	5.6
101–200	55	0	30	85	1	4.5	55	0	30	85	1	4.5	55	0	30	85	1	4.5
201–300	60	0	30	90	1	4.3	60	0	30	90	1	4.3	60	0	30	90	1	4.3
301–400	70	0	70	140	2	3.8	70	0	70	140	2	3.8	70	0	70	140	2	3.8
401–500	70	0	75	145	2	3.7	70	0	75	145	2	3.7	70	0	75	145	2	3.7
501–600	75	0	80	155	2	3.6	75	0	80	155	2	3.6	80	0	120	200	3	3.3
601–800	75	0	85	160	2	3.5	85	0	130	215	3	3.2	85	0	130	215	3	3.2
801–1000	85	0	140	225	3	3.1	85	0	140	225	3	3.1	95	0	185	280	4	2.9
1001–2000	95	0	155	250	3	2.9	100	0	215	315	4	2.7	175	1	265	440	6	2.5
2001–3000	105	0	230	335	4	2.6	185	1	280	465	6	2.4	205	1	405	610	8	2.2
3001–4000	110	0	290	400	5	2.5	195	1	345	540	7	2.3	285	2	475	760	10	2.1
4001–5000	195	1	300	495	6	2.3	210	1	415	625	8	2.2	300	2	560	860	11	2.0
5001–7000	200	1	375	575	7	2.2	300	2	510	810	10	2.0	390	3	710	1100	14	1.9
7001–10,000	205	1	375	580	7	2.2	320	2	665	985	12	1.9	415	3	955	1370	17	1.8
10,001–20,000	215	1	455	670	8	2.1	335	2	855	1190	14	1.8	600	5	1200	1800	22	1.7
20,001–50,000	225	1	605	830	10	2.0	440	3	1030	1470	17	1.7	800	7	1640	2440	29	1.6
50,001–100,000	335	2	725	1060	12	1.8	555	4	1305	1860	21	1.6	1025	9	2405	3430	40	1.5

Trial 1: n_1 = first sample size; c_1 = acceptance number for first sample
"All" indicates that each piece in the lot is to be inspected
Trial 2: n_2 = second sample size; c_2 = acceptance number for first and second samples combined
p_t = lot tolerance per cent defective with a Consumer's Risk (P_C) of 0.10

Double Sampling Table for Average Outgoing Quality Limit (AOQL) = 1.0%

Lot Size	Process Average 0 to 0.02% Trial 1 n_1 c_1	Trial 2 n_2 n_1+n_2 c_2	p_t %	Process Average 0.03 to 0.20% Trial 1 n_1 c_1	Trial 2 n_2 n_1+n_2 c_2	p_t %	Process Average 0.21 to 0.40% Trial 1 n_1 c_1	Trial 2 n_2 n_1+n_2 c_2	p_t %
1–25	All 0	— — —	—	All 0	— — —	—	All 0	— — —	—
26–50	22 0	— — —	7.7	22 0	— — —	7.7	22 0	— — —	7.7
51–100	33 0	17 50 1	6.9	33 0	17 50 1	6.9	33 0	17 50 1	6.9
101–200	43 0	22 65 1	5.8	43 0	22 65 1	5.8	43 0	22 65 1	5.8
201–300	47 0	28 75 1	5.5	47 0	28 75 1	5.5	47 0	28 75 1	5.5
301–400	49 0	31 80 1	5.4	49 0	31 80 1	5.4	55 0	60 115 2	4.8
401–500	50 0	30 80 1	5.4	50 0	30 80 1	5.4	55 0	65 120 2	4.7
501–600	50 0	30 80 1	5.4	50 0	30 80 1	5.4	60 0	65 125 2	4.6
601–800	50 0	35 85 1	5.3	60 0	70 130 2	4.5	60 0	70 130 2	4.5
801–1000	55 0	30 85 1	5.2	60 0	75 135 2	4.4	60 0	75 135 2	4.4
1001–2000	55 0	35 90 1	5.1	65 0	75 140 2	4.3	75 0	120 195 3	3.8
2001–3000	65 0	80 145 2	4.2	65 0	80 145 2	4.2	75 0	125 200 3	3.7
3001–4000	70 0	80 150 2	4.1	70 0	80 150 2	4.1	80 0	175 255 4	3.5
4001–5000	70 0	80 150 2	4.1	70 0	80 150 2	4.1	80 0	180 260 4	3.4
5001–7000	70 0	80 150 2	4.1	75 0	125 200 3	3.7	80 0	180 260 4	3.4
7001–10,000	70 0	80 150 2	4.1	80 0	125 205 3	3.6	85 0	180 265 4	3.3
10,001–20,000	70 0	80 150 2	4.1	80 0	130 210 3	3.6	90 0	230 320 5	3.2
20,001–50,000	75 0	80 155 2	4.0	80 0	135 215 3	3.6	95 0	300 395 6	2.9
50,001–100,000	75 0	80 155 2	4.0	85 0	180 265 4	3.3	170 1	380 550 8	2.6

Lot Size	Process Average 0.41 to 0.60% Trial 1 n_1 c_1	Trial 2 n_2 n_1+n_2 c_2	p_t %	Process Average 0.61 to 0.80% Trial 1 n_1 c_1	Trial 2 n_2 n_1+n_2 c_2	p_t %	Process Average 0.81 to 1.00% Trial 1 n_1 c_1	Trial 2 n_2 n_1+n_2 c_2	p_t %
1–25	All 0	— — —	—	All 0	— — —	—	All 0	— — —	—
26–50	22 0	— — —	7.7	22 0	— — —	7.7	22 0	— — —	7.7
51–100	33 0	17 50 1	6.9	33 0	17 50 1	6.9	33 0	17 50 1	6.9
101–200	43 0	22 65 1	5.8	43 0	22 65 1	5.8	47 0	43 90 2	5.4
201–300	55 0	50 105 2	4.9	55 0	50 105 2	4.9	55 0	50 105 2	4.9
301–400	55 0	60 115 2	4.8	55 0	60 115 2	4.8	60 0	80 140 3	4.5
401–500	55 0	65 120 2	4.7	60 0	95 155 3	4.3	60 0	95 155 3	4.3
501–600	60 0	65 125 2	4.6	65 0	100 165 3	4.2	65 0	100 165 3	4.2
601–800	65 0	105 170 3	4.1	65 0	105 170 3	4.1	70 0	140 210 4	3.9
801–1000	65 0	110 175 3	4.0	70 0	150 220 4	3.8	125 1	180 305 6	3.5
1001–2000	80 0	165 245 4	3.7	135 1	200 335 6	3.3	140 1	245 385 7	3.2
2001–3000	80 0	170 250 4	3.6	150 1	265 415 7	3.0	215 2	355 570 10	2.8
3001–4000	85 0	220 305 5	3.3	160 1	330 490 8	2.8	225 2	455 680 12	2.7
4001–5000	145 1	225 370 6	3.1	225 2	375 600 10	2.7	240 2	595 835 14	2.5
5001–7000	155 1	285 440 7	2.9	235 2	440 675 11	2.6	310 3	665 975 16	2.4
7001–10,000	165 1	355 520 8	2.7	250 2	585 835 13	2.4	385 4	785 1170 19	2.3
10,001–20,000	175 1	415 590 9	2.6	325 3	655 980 15	2.3	520 6	980 1500 24	2.2
20,001–50,000	250 2	490 740 11	2.4	340 3	910 1250 19	2.2	610 7	1410 2020 32	2.1
50,001–100,000	275 2	700 975 14	2.2	420 4	1050 1470 22	2.1	770 9	1850 2620 41	2.0

Trial 1: n_1 = first sample size; c_1 = acceptance number for first sample
"All" indicates that each piece in the lot is to be inspected
Trial 2: n_2 = second sample size; c_2 = acceptance number for first and second samples combined
p_t = lot tolerance per cent defective with a Consumer's Risk (P_C) of 0.10

Double Sampling Table for
Average Outgoing Quality Limit (AOQL) = 1.5%

Lot Size	Process Average 0 to 0.03% — Trial 1 n_1	c_1	Trial 2 n_2	n_1+n_2	c_2	p_t %	Process Average 0.04 to 0.30% — Trial 1 n_1	c_1	Trial 2 n_2	n_1+n_2	c_2	p_t %	Process Average 0.31 to 0.60% — Trial 1 n_1	c_1	Trial 2 n_2	n_1+n_2	c_2	p_t %
1–15	All	0	–	–	–	–	All	0	–	–	–	–	All	0	–	–	–	–
16–50	16	0	–	–	–	11.6	16	0	–	–	–	11.6	16	0	–	–	–	11.6
51–75	23	0	11	34	1	10.5	23	0	11	34	1	10.5	23	0	11	34	1	10.5
76–100	26	0	14	40	1	9.4	26	0	14	40	1	9.4	26	0	14	40	1	9.4
101–200	31	0	18	49	1	8.4	31	0	18	49	1	8.4	31	0	18	49	1	8.4
201–300	33	0	22	55	1	8.0	33	0	22	55	1	8.0	38	0	37	75	2	7.0
301–400	34	0	21	55	1	7.9	34	0	21	55	1	7.9	39	0	41	80	2	6.9
401–500	35	0	20	55	1	7.8	35	0	20	55	1	7.8	39	0	46	85	2	6.9
501–600	35	0	20	55	1	7.8	40	0	45	85	2	6.8	40	0	45	85	2	6.8
601–800	35	0	20	55	1	7.8	41	0	49	90	2	6.7	46	0	74	120	3	6.0
801–1000	36	0	19	55	1	7.8	42	0	48	90	2	6.5	47	0	78	125	3	5.9
1001–2000	44	0	51	95	2	6.3	44	0	51	95	2	6.3	49	0	81	130	3	5.7
2001–3000	45	0	50	95	2	6.2	45	0	50	95	2	6.2	55	0	110	165	4	5.3
3001–4000	45	0	50	95	2	6.2	50	0	85	135	3	5.5	55	0	115	170	4	5.2
4001–5000	45	0	50	95	2	6.2	50	0	85	135	3	5.5	55	0	120	175	4	5.1
5001–7000	46	0	54	100	2	6.1	50	0	90	140	3	5.4	60	0	155	215	5	4.7
7001–10,000	46	0	54	100	2	6.1	50	0	90	140	3	5.4	60	0	160	220	5	4.6
10,001–20,000	46	0	54	100	2	6.1	50	0	90	140	3	5.4	60	0	165	225	5	4.5
20,001–50,000	47	0	53	100	2	6.1	55	0	125	180	4	5.0	65	0	195	260	6	4.4
50,001–100,000	47	0	53	100	2	6.1	55	0	130	185	4	4.9	115	1	235	350	8	4.0

Lot Size	Process Average 0.61 to 0.90% — Trial 1 n_1	c_1	Trial 2 n_2	n_1+n_2	c_2	p_t %	Process Average 0.91 to 1.20% — Trial 1 n_1	c_1	Trial 2 n_2	n_1+n_2	c_2	p_t %	Process Average 1.21 to 1.50% — Trial 1 n_1	c_1	Trial 2 n_2	n_1+n_2	c_2	p_t %
1–15	All	0	–	–	–	–	All	0	–	–	–	–	All	0	–	–	–	–
16–50	16	0	–	–	–	11.6	16	0	–	–	–	11.6	16	0	–	–	–	11.6
51–75	23	0	11	34	1	10.5	23	0	11	34	1	10.5	23	0	11	34	1	10.5
76–100	26	0	14	40	1	9.4	26	0	14	40	1	9.4	26	0	14	40	1	9.4
101–200	35	0	35	70	2	7.5	35	0	35	70	2	7.5	35	0	35	70	2	7.5
201–300	38	0	37	75	2	7.0	38	0	37	75	2	7.0	40	0	60	100	3	6.6
301–400	39	0	41	80	2	6.9	42	0	63	105	3	6.3	42	0	63	105	3	6.3
401–500	44	0	71	115	3	6.1	44	0	71	115	3	6.1	46	0	94	140	4	5.8
501–600	44	0	71	115	3	6.1	44	0	71	115	3	6.1	47	0	98	145	4	5.7
601–800	46	0	74	120	3	6.0	49	0	101	150	4	5.6	85	1	125	210	6	5.0
801–1000	47	0	78	125	3	5.9	50	0	105	155	4	5.5	90	1	125	215	6	4.9
1001–2000	55	0	105	160	4	5.4	95	1	175	270	7	4.6	100	1	230	330	9	4.4
2001–3000	100	1	145	245	6	4.6	110	1	255	365	9	4.1	155	2	345	500	13	3.9
3001–4000	105	1	190	295	7	4.3	160	2	300	460	11	3.8	205	3	405	610	15	3.6
4001–5000	105	1	190	295	7	4.3	165	2	340	505	12	3.7	250	4	480	730	18	3.5
5001–7000	110	1	225	335	8	4.2	165	2	375	540	13	3.7	310	5	610	920	22	3.3
7001–10,000	115	1	280	395	9	3.9	170	2	420	590	14	3.6	360	6	660	1020	24	3.2
10,001–20,000	120	1	315	435	10	3.8	210	3	420	630	15	3.6	415	7	835	1250	29	3.1
20,001–50,000	165	2	350	515	12	3.7	225	3	640	865	20	3.3	510	9	1130	1640	38	3.0
50,001–100,000	175	2	440	615	14	3.5	275	4	725	1000	23	3.2	570	10	1400	1970	45	2.9

Trial 1: n_1 = first sample size; c_1 = acceptance number for first sample
"All" indicates that each piece in the lot is to be inspected
Trial 2: n_2 = second sample size; c_2 = acceptance number for first and second samples combined
p_t = lot tolerance per cent defective with a Consumer's Risk (P_C) of 0.10

DOUBLE SAMPLING

2.0%

AOQL

Double Sampling Table for
Average Outgoing Quality Limit (AOQL) = 2.0%

Lot Size	Process Average 0 to 0.04%					Process Average 0.05 to 0.40%					Process Average 0.41 to 0.80%				
	Trial 1		Trial 2			p_t %	Trial 1		Trial 2			p_t %	Trial 1		Trial 2			p_t %
	n_1	c_1	n_2	n_1+n_2	c_2		n_1	c_1	n_2	n_1+n_2	c_2		n_1	c_1	n_2	n_1+n_2	c_2	
1–15	All	0	—	—	—	—	All	0	—	—	—	—	All	0	—	—	—	—
16–50	14	0	—	—	—	13.6	14	0	—	—	—	13.6	14	0	—	—	—	13.6
51–100	21	0	12	33	1	11.7	21	0	12	33	1	11.7	21	0	12	33	1	11.7
101–200	24	0	13	37	1	11.0	24	0	13	37	1	11.0	24	0	13	37	1	11.0
201–300	26	0	15	41	1	10.4	26	0	15	41	1	10.4	29	0	31	60	2	9.1
301–400	26	0	16	42	1	10.3	26	0	16	42	1	10.3	30	0	35	65	2	9.0
401–500	27	0	16	43	1	10.3	30	0	35	65	2	9.0	30	0	35	65	2	9.0
501–600	27	0	16	43	1	10.3	31	0	34	65	2	8.9	35	0	55	90	3	7.9
601–800	27	0	17	44	1	10.2	31	0	39	70	2	8.8	35	0	60	95	3	7.7
801–1000	27	0	17	44	1	10.2	32	0	38	70	2	8.7	36	0	59	95	3	7.6
1001–2000	33	0	37	70	2	8.5	33	0	37	70	2	8.5	37	0	63	100	3	7.5
2001–3000	34	0	41	75	2	8.2	34	0	41	75	2	8.2	41	0	84	125	4	7.0
3001–4000	34	0	41	75	2	8.2	38	0	62	100	3	7.3	41	0	89	130	4	6.9
4001–5000	34	0	41	75	2	8.2	38	0	62	100	3	7.3	42	0	88	130	4	6.9
5001–7000	35	0	40	75	2	8.1	38	0	62	100	3	7.3	44	0	116	160	5	6.4
7001–10,000	35	0	40	75	2	8.1	38	0	62	100	3	7.3	45	0	115	160	5	6.3
10,001–20,000	35	0	40	75	2	8.1	39	0	66	105	3	7.2	45	0	115	160	5	6.3
20,001–50,000	35	0	40	75	2	8.1	43	0	92	135	4	6.6	47	0	148	195	6	6.0
50,001–100,000	35	0	45	80	2	8.0	43	0	92	135	4	6.6	85	1	185	270	8	5.2

Lot Size	Process Average 0.81 to 1.20%					Process Average 1.21 to 1.60%					Process Average 1.61 to 2.00%				
	Trial 1		Trial 2			p_t %	Trial 1		Trial 2			p_t %	Trial 1		Trial 2			p_t %
	n_1	c_1	n_2	n_1+n_2	c_2		n_1	c_1	n_2	n_1+n_2	c_2		n_1	c_1	n_2	n_1+n_2	c_2	
1–15	All	0	—	—	—	—	All	0	—	—	—	—	All	0	—	—	—	—
16–50	14	0	—	—	—	13.6	14	0	—	—	—	13.6	23	0	23	46	2	10.9
51–100	21	0	12	33	1	11.7	21	0	12	33	1	11.7	27	0	28	55	2	9.6
101–200	27	0	28	55	2	9.6	27	0	28	55	2	9.6	27	0	28	55	2	9.6
201–300	29	0	31	60	2	9.1	32	0	48	80	3	8.4	32	0	48	80	3	8.4
301–400	33	0	52	85	3	8.2	33	0	52	85	3	8.2	36	0	69	105	4	7.6
401–500	34	0	56	90	3	7.9	36	0	74	110	4	7.5	60	1	90	150	6	7.0
501–600	35	0	55	90	3	7.9	37	0	78	115	4	7.4	65	1	95	160	6	6.8
601–800	38	0	82	120	4	7.3	38	0	82	120	4	7.3	70	1	120	190	7	6.4
801–1000	38	0	87	125	4	7.2	70	1	100	170	6	6.5	70	1	145	215	8	6.2
1001–2000	43	0	112	155	5	6.5	80	1	160	240	8	5.8	110	2	205	315	11	5.5
2001–3000	75	1	115	190	6	6.1	115	2	195	310	10	5.3	160	3	310	470	15	4.7
3001–4000	80	1	140	220	7	5.8	120	2	255	375	12	5.0	235	5	415	650	20	4.3
4001–5000	80	1	175	255	8	5.5	125	2	285	410	13	4.9	275	6	475	750	23	4.2
5001–7000	85	1	205	290	9	5.3	125	2	320	445	14	4.8	280	6	575	855	26	4.1
7001–10,000	85	1	210	295	9	5.2	165	3	335	500	15	4.5	320	7	645	965	29	4.0
10,001–20,000	90	1	260	350	11	5.1	170	3	425	595	18	4.4	395	9	835	1230	37	3.9
20,001–50,000	130	2	300	430	13	4.7	205	4	515	720	22	4.3	480	11	1090	1570	46	3.7
50,001–100,000	135	2	345	480	14	4.5	250	5	615	865	26	4.1	580	13	1460	2040	58	3.5

Trial 1: n_1 = first sample size; c_1 = acceptance number for first sample
"All" indicates that each piece in the lot is to be inspected
Trial 2: n_2 = second sample size; c_2 = acceptance number for first and second samples combined
p_t = lot tolerance per cent defective with a Consumer's Risk (P_C) of 0.10

Double Sampling Table for
Average Outgoing Quality Limit (AOQL) = 2.5%

Lot Size	Process Average 0 to 0.05% Trial 1 n_1	c_1	Trial 2 n_2	n_1+n_2	c_2	p_t %	Process Average 0.06 to 0.50% Trial 1 n_1	c_1	Trial 2 n_2	n_1+n_2	c_2	p_t %	Process Average 0.51 to 1.00% Trial 1 n_1	c_1	Trial 2 n_2	n_1+n_2	c_2	p_t %
1–10	All	0	–	–	–	–	All	0	–	–	–	–	All	0	–	–	–	–
11–50	11	0	–	–	–	17.6	11	0	–	–	–	17.6	11	0	–	–	–	17.6
51–100	18	0	10	28	1	14.1	18	0	10	28	1	14.1	18	0	10	28	1	14.1
101–200	20	0	11	31	1	13.7	20	0	11	31	1	13.7	23	0	25	48	2	11.7
201–300	21	0	13	34	1	13.0	21	0	13	34	1	13.0	24	0	25	49	2	11.4
301–400	21	0	14	35	1	12.8	24	0	26	50	2	11.3	24	0	26	50	2	11.3
401–500	22	0	13	35	1	12.7	25	0	25	50	2	11.1	28	0	47	75	3	9.8
501–600	22	0	14	36	1	12.5	25	0	30	55	2	10.9	28	0	47	75	3	9.8
601–800	22	0	14	36	1	12.5	26	0	29	55	2	10.8	28	0	47	75	3	9.8
801–1000	26	0	29	55	2	10.8	26	0	29	55	2	10.8	29	0	46	75	3	9.6
1001–2000	27	0	33	60	2	10.5	27	0	33	60	2	10.5	33	0	72	105	4	8.3
2001–3000	27	0	33	60	2	10.5	30	0	50	80	3	9.3	33	0	72	105	4	8.3
3001–4000	27	0	33	60	2	10.5	31	0	49	80	3	9.1	33	0	77	110	4	8.2
4001–5000	27	0	33	60	2	10.5	31	0	49	80	3	9.1	36	0	94	130	5	7.6
5001–7000	28	0	32	60	2	10.3	31	0	49	80	3	9.1	36	0	94	130	5	7.7
7001–10,000	28	0	32	60	2	10.3	31	0	49	80	3	9.2	36	0	94	130	5	7.7
10,001–20,000	28	0	32	60	2	10.3	31	0	49	80	3	9.2	36	0	94	130	5	7.8
20,001–50,000	28	0	32	60	2	10.3	33	0	87	120	4	7.7	70	1	145	215	8	6.6
50,001–100,000	28	0	37	65	2	10.2	33	0	92	125	4	7.6	70	1	170	240	9	6.4

Lot Size	Process Average 1.01 to 1.50% Trial 1 n_1	c_1	Trial 2 n_2	n_1+n_2	c_2	p_t %	Process Average 1.51 to 2.00% Trial 1 n_1	c_1	Trial 2 n_2	n_1+n_2	c_2	p_t %	Process Average 2.01 to 2.50% Trial 1 n_1	c_1	Trial 2 n_2	n_1+n_2	c_2	p_t %
1–10	All	0	–	–	–	–	All	0	–	–	–	–	All	0	–	–	–	–
11–50	11	0	–	–	–	17.6	11	0	–	–	–	17.6	11	0	–	–	–	17.6
51–100	18	0	10	28	1	14.1	20	0	20	40	2	13.0	20	0	20	40	2	13.0
101–200	23	0	25	48	2	11.7	23	0	25	48	2	11.7	25	0	35	60	3	10.8
201–300	26	0	44	70	3	10.3	26	0	44	70	3	10.3	28	0	57	85	4	9.5
301–400	27	0	43	70	3	9.9	29	0	61	90	4	9.3	49	1	71	120	6	8.8
401–500	28	0	47	75	3	9.8	30	0	60	90	4	9.2	50	1	80	130	6	8.4
501–600	30	0	65	95	4	9.1	30	0	65	95	4	9.1	55	1	95	150	7	8.0
601–800	31	0	69	100	4	8.8	55	1	85	140	6	8.0	60	1	115	175	8	7.6
801–1000	32	0	68	100	4	8.7	60	1	100	160	7	7.8	85	2	120	205	9	7.2
1001–2000	60	1	90	150	6	7.6	65	1	150	215	9	7.0	95	2	210	305	13	6.5
2001–3000	65	1	115	180	7	7.2	90	2	170	260	11	6.8	125	3	265	390	16	6.0
3001–4000	65	1	140	205	8	6.8	95	2	205	300	12	6.4	185	5	350	535	21	5.5
4001–5000	70	1	160	230	9	6.5	100	2	255	355	14	6.0	220	6	410	630	24	5.2
5001–7000	75	1	190	265	10	6.2	130	3	265	395	15	5.7	255	7	495	750	28	5.0
7001–10,000	100	2	195	295	11	6.0	140	3	355	495	18	5.3	325	9	665	990	36	4.7
10,001–20,000	105	2	215	320	12	5.9	170	4	380	550	20	5.2	360	10	830	1190	43	4.6
20,001 50,000	105	2	245	350	13	5.8	205	5	485	690	25	5.0	415	11	1145	1560	54	4.3
50,001–100,000	110	2	295	405	15	5.6	245	6	610	855	30	4.7	510	14	1370	1880	65	4.2

Trial 1: n_1 = first sample size; c_1 = acceptance number for first sample
"All" indicates that each piece in the lot is to be inspected
Trial 2: n_2 = second sample size; c_2 = acceptance number for first and second samples combined
p_t = lot tolerance per cent defective with a Consumer's Risk (P_C) of 0.10

3.0% AOQL

Double Sampling Table for Average Outgoing Quality Limit (AOQL) = 3.0%

Lot Size	Process Average 0 to 0.06% Trial 1 n_1	c_1	Trial 2 n_2	n_1+n_2	c_2	p_t %	Process Average 0.07 to 0.60% Trial 1 n_1	c_1	Trial 2 n_2	n_1+n_2	c_2	p_t %	Process Average 0.61 to 1.20% Trial 1 n_1	c_1	Trial 2 n_2	n_1+n_2	c_2	p_t %
1–10	All	0	–	–	–	–	All	0	–	–	–	–	All	0	–	–	–	–
11–50	10	0	–	–	–	19.0	10	0	–	–	–	19.0	10	0	–	–	–	19.0
51–100	16	0	9	25	1	16.4	16	0	9	25	1	16.4	16	0	9	25	1	16.4
101–200	17	0	9	26	1	16.0	17	0	9	26	1	16.0	17	0	9	26	1	16.0
201–300	18	0	10	28	1	15.5	18	0	10	28	1	15.5	21	0	23	44	2	13.3
301–400	18	0	11	29	1	15.2	21	0	24	45	2	13.2	23	0	37	60	3	12.0
401–500	18	0	11	29	1	15.2	21	0	25	46	2	13.0	24	0	36	60	3	11.7
501–600	18	0	12	30	1	15.0	21	0	25	46	2	13.0	24	0	41	65	3	11.5
601–800	21	0	25	46	2	13.0	21	0	25	46	2	13.0	24	0	41	65	3	11.5
801–1000	21	0	26	47	2	12.8	21	0	26	47	2	12.8	25	0	40	65	3	11.4
1001–2000	22	0	26	48	2	12.6	22	0	26	48	2	12.6	27	0	58	85	4	10.3
2001–3000	22	0	26	48	2	12.6	25	0	40	65	3	11.4	28	0	62	90	4	10.0
3001–4000	23	0	26	49	2	12.4	25	0	45	70	3	11.0	29	0	76	105	5	9.6
4001–5000	23	0	26	49	2	12.4	26	0	44	70	3	11.0	30	0	75	105	5	9.5
5001–7000	23	0	27	50	2	12.2	26	0	44	70	3	11.0	30	0	80	110	5	9.4
7001–10,000	23	0	27	50	2	12.2	27	0	43	70	3	11.0	30	0	80	110	5	9.4
10,001–20,000	23	0	27	50	2	12.2	27	0	43	70	3	11.0	31	0	94	125	6	9.2
20,001–50,000	23	0	27	50	2	12.2	28	0	67	95	4	9.7	55	1	120	175	8	8.0
50,001–100,000	23	0	27	50	2	12.2	31	0	84	115	5	9.0	60	1	140	200	9	7.6

Lot Size	Process Average 1.21 to 1.80% Trial 1 n_1	c_1	Trial 2 n_2	n_1+n_2	c_2	p_t %	Process Average 1.81 to 2.40% Trial 1 n_1	c_1	Trial 2 n_2	n_1+n_2	c_2	p_t %	Process Average 2.41 to 3.00% Trial 1 n_1	c_1	Trial 2 n_2	n_1+n_2	c_2	p_t %
1–10	All	0	–	–	–	–	All	0	–	–	–	–	All	0	–	–	–	–
11–50	10	0	–	–	–	19.0	10	0	–	–	–	19.0	10	0	–	–	–	19.0
51–100	17	0	17	34	2	15.8	17	0	17	34	2	15.8	17	0	17	34	2	15.8
101–200	20	0	21	41	2	13.7	22	0	33	55	3	12.4	22	0	33	55	3	12.4
201–300	23	0	37	60	3	12.0	23	0	37	60	3	12.0	24	0	51	75	4	11.1
301–400	23	0	37	60	3	12.0	25	0	55	80	4	10.8	42	1	63	105	6	10.4
401–500	24	0	36	60	3	11.7	25	0	55	80	4	10.8	46	1	79	125	7	9.7
501–600	26	0	54	80	4	10.7	46	1	69	115	6	9.7	48	1	97	145	8	9.2
601–800	26	0	54	80	4	10.7	49	1	81	130	7	9.4	50	1	115	165	9	8.9
801–1000	27	0	58	85	4	10.3	49	1	86	135	7	9.2	70	2	120	190	10	8.4
1001–2000	49	1	76	125	6	9.1	50	1	150	200	10	8.0	100	3	180	280	14	7.5
2001–3000	50	1	95	145	7	8.7	80	2	165	245	12	7.6	130	4	260	390	19	6.9
3001–4000	55	1	110	165	8	8.5	105	3	200	305	14	7.0	155	5	330	485	23	6.5
4001–5000	60	1	135	195	9	7.8	110	3	225	335	15	6.7	215	7	390	605	27	6.0
5001–7000	60	1	165	225	10	7.3	110	3	250	360	16	6.6	270	9	505	775	34	5.7
7001–10,000	85	2	160	245	11	7.2	115	3	290	405	18	6.5	285	9	680	965	41	5.4
10,001–20,000	85	2	180	265	12	7.2	140	4	315	455	20	6.3	315	10	805	1120	47	5.3
20,001–50,000	85	2	205	290	13	7.0	170	5	420	590	26	6.0	390	13	940	1330	56	5.2
50,001–100,000	90	2	245	335	15	6.8	200	6	505	705	30	5.7	445	15	1105	1550	65	5.1

Trial 1: n_1 = first sample size; c_1 = acceptance number for first sample
"All" indicates that each piece in the lot is to be inspected
Trial 2: n_2 = second sample size; c_2 = acceptance number for first and second samples combined
p_t = lot tolerance per cent defective with a Consumer's Risk (P_C) of 0.10

DOUBLE SAMPLING

4.0%
AOQL

Double Sampling Table for
Average Outgoing Quality Limit (AOQL) = 4.0%

Lot Size	Process Average 0 to 0.08%						Process Average 0.09 to 0.80%						Process Average 0.81 to 1.60%					
	Trial 1		Trial 2			p_t %	Trial 1		Trial 2			p_t %	Trial 1		Trial 2			p_t %
	n_1	c_1	n_2	n_1+n_2	c_2		n_1	c_1	n_2	n_1+n_2	c_2		n_1	c_1	n_2	n_1+n_2	c_2	
1–10	All	0	–	–	–	–	All	0	–	–	–	–	All	0	–	–	–	–
11–50	8	0	–	–	–	23.0	8	0	–	–	–	23.0	8	0	–	–	–	23.0
51–100	12	0	7	19	1	22.0	12	0	7	19	1	22.0	12	0	7	19	1	22.0
101–200	13	0	8	21	1	21.0	13	0	8	21	1	21.0	15	0	17	32	2	18.0
201–300	13	0	9	22	1	20.5	16	0	18	34	2	17.4	16	0	18	34	2	17.4
301–400	14	0	8	22	1	20.0	16	0	19	35	2	17.0	18	0	28	46	3	15.5
401–500	14	0	8	22	1	20.0	16	0	19	35	2	17.0	19	0	28	47	3	15.3
501–600	16	0	19	35	2	17.0	16	0	19	35	2	17.0	19	0	29	48	3	15.1
601–800	16	0	20	36	2	16.7	16	0	20	36	2	16.7	19	0	30	49	3	14.9
801–1000	16	0	20	36	2	16.7	16	0	20	36	2	16.7	20	0	45	65	4	13.8
1001–2000	17	0	19	36	2	16.6	19	0	31	50	3	14.8	21	0	44	65	4	13.6
2001–3000	17	0	19	36	2	16.6	19	0	31	50	3	14.8	21	0	44	65	4	13.6
3001–4000	17	0	20	37	2	16.5	19	0	31	50	3	14.8	22	0	58	80	5	13.0
4001–5000	17	0	20	37	2	16.5	19	0	31	50	3	14.8	22	0	58	80	5	13.0
5001–7000	17	0	20	37	2	16.5	19	0	31	50	3	14.8	22	0	58	80	5	13.0
7001–10,000	17	0	20	37	2	16.5	19	0	36	55	3	14.6	23	0	57	80	5	12.7
10,001–20,000	17	0	20	37	2	16.5	21	0	44	65	4	13.6	23	0	72	95	6	12.0
20,001–50,000	17	0	20	37	2	16.5	21	0	44	65	4	13.6	43	1	92	135	8	10.6
50,001–100,000	17	0	20	37	2	16.5	23	0	62	85	5	12.5	44	1	106	150	9	10.3

Lot Size	Process Average 1.61 to 2.40%						Process Average 2.41 to 3.20%						Process Average 3.21 to 4.00%					
	Trial 1		Trial 2			p_t %	Trial 1		Trial 2			p_t %	Trial 1		Trial 2			p_t %
	n_1	c_1	n_2	n_1+n_2	c_2		n_1	c_1	n_2	n_1+n_2	c_2		n_1	c_1	n_2	n_1+n_2	c_2	
1–10	All	0	–	–	–	–	All	0	–	–	–	–	All	0	–	–	–	–
11–50	8	0	–	–	–	23.0	8	0	–	–	–	23.0	8	0	–	–	–	23.0
51–100	13	0	14	27	2	20.5	13	0	14	27	2	20.5	13	0	14	27	2	20.5
101–200	16	0	26	42	3	16.5	16	0	26	42	3	16.5	16	0	26	42	3	16.5
201–300	17	0	28	45	3	16.0	18	0	37	55	4	15.0	33	1	47	80	6	13.2
301–400	19	0	41	60	4	14.3	19	0	41	60	4	14.3	35	1	60	95	7	12.8
401–500	20	0	40	60	4	14.0	34	1	51	85	6	13.0	36	1	74	110	8	12.2
501–600	20	0	45	65	4	13.8	37	1	63	100	7	12.2	50	2	75	125	9	11.6
601–800	22	0	58	80	5	13.0	39	1	81	120	8	11.6	55	2	105	160	11	10.8
801–1000	37	1	58	95	6	12.2	41	1	94	135	9	11.1	55	2	120	175	12	10.5
1001–2000	39	1	71	110	7	11.5	55	2	110	165	11	10.6	80	3	165	245	16	9.5
2001–3000	41	1	89	130	8	11.0	60	2	145	205	13	9.8	95	4	210	305	20	9.2
3001–4000	43	1	102	145	9	10.5	80	3	160	240	15	9.4	115	5	250	365	23	8.8
4001–5000	45	1	120	165	10	10.0	85	3	180	265	16	8.9	160	7	305	465	28	8.1
5001–7000	65	2	120	185	11	9.6	85	3	200	285	17	8.7	210	9	450	660	38	7.4
7001–10,000	65	2	140	205	12	9.3	90	3	230	320	19	8.5	235	10	550	785	44	7.1
10,001–20,000	65	2	160	225	13	9.0	105	4	265	370	22	8.3	270	12	625	895	50	7.0
20,001–50,000	70	2	175	245	14	8.8	125	5	315	440	26	8.1	295	13	725	1020	57	6.9
50,001–100,000	70	2	205	275	16	8.7	150	6	385	535	31	7.7	335	15	845	1180	66	6.8

Trial 1: n_1 = first sample size; c_1 = acceptance number for first sample
"All" indicates that each piece in the lot is to be inspected
Trial 2: n_2 = second sample size; c_2 = acceptance number for first and second samples combined
p_t = lot tolerance per cent defective with a Consumer's Risk (P_C) of 0.10

5.0%
AOQL

Double Sampling Table for Average Outgoing Quality Limit (AOQL) = 5.0%

Lot Size	Process Average 0 to 0.10%						Process Average 0.11 to 1.00%						Process Average 1.01 to 2.00%					
	Trial 1		Trial 2			p_t %	Trial 1		Trial 2			p_t %	Trial 1		Trial 2			p_t %
	n_1	c_1	n_2	n_1+n_2	c_2		n_1	c_1	n_2	n_1+n_2	c_2		n_1	c_1	n_2	n_1+n_2	c_2	
1–5	All	0	–	–	–	–	All	0	–	–	–	–	All	0	–	–	–	–
6–50	6	0	–	–	–	30.5	6	0	–	–	–	30.5	6	0	–	–	–	30.5
51–100	10	0	6	16	1	26.5	10	0	6	16	1	26.5	11	0	11	22	2	25.0
101–200	11	0	6	17	1	26.0	12	0	15	27	2	22.0	12	0	15	27	2	22.0
201–300	11	0	7	18	1	25.0	13	0	15	28	2	21.0	14	0	24	38	3	19.3
301–400	11	0	8	19	1	25.0	13	0	15	28	2	21.0	15	0	24	39	3	19.0
401–500	13	0	15	28	2	21.0	13	0	15	28	2	21.0	15	0	24	39	3	19.0
501–600	13	0	15	28	2	21.0	13	0	15	28	2	21.0	15	0	25	40	3	18.7
601–800	13	0	16	29	2	20.5	13	0	16	29	2	20.5	16	0	34	50	4	17.1
801–1000	13	0	16	29	2	20.5	13	0	16	29	2	20.5	16	0	34	50	4	17.1
1001–2000	13	0	16	29	2	20.5	15	0	25	40	3	18.7	17	0	33	50	4	17.1
2001–3000	13	0	16	29	2	21.0	15	0	26	41	3	18.4	17	0	48	65	5	15.5
3001–4000	14	0	15	29	2	21.0	15	0	26	41	3	18.4	18	0	47	65	5	15.5
4001–5000	14	0	16	30	2	20.5	16	0	25	41	3	18.0	18	0	47	65	5	15.5
5001–7000	14	0	16	30	2	20.5	16	0	26	42	3	18.0	18	0	47	65	5	15.5
7001–10,000	14	0	16	30	2	20.5	16	0	26	42	3	18.0	19	0	56	75	6	15.0
10,001–20,000	14	0	17	31	2	20.5	17	0	38	55	4	16.4	19	0	56	75	6	15.0
20,001–50,000	14	0	17	31	2	20.5	17	0	38	55	4	16.4	33	1	72	105	8	13.5
50,001–100,000	14	0	18	32	2	20.5	18	0	47	65	5	15.6	34	1	86	120	9	13.1

Lot Size	Process Average 2.01 to 3.00%						Process Average 3.01 to 4.00%						Process Average 4.01 to 5.00%					
	Trial 1		Trial 2			p_t %	Trial 1		Trial 2			p_t %	Trial 1		Trial 2			p_t %
	n_1	c_1	n_2	n_1+n_2	c_2		n_1	c_1	n_2	n_1+n_2	c_2		n_1	c_1	n_2	n_1+n_2	c_2	
1–5	All	0	–	–	–	–	All	0	–	–	–	–	All	0	–	–	–	–
6–50	6	0	–	–	–	30.5	6	0	–	–	–	30.5	6	0	–	–	–	30.5
51–100	11	0	11	22	2	25.0	11	0	11	22	2	25.0	12	0	18	30	3	23.0
101–200	14	0	22	36	3	19.8	14	0	22	36	3	19.8	14	0	30	44	4	19.0
201–300	14	0	24	38	3	19.3	15	0	32	47	4	18.0	27	1	48	75	7	16.3
301–400	16	0	33	49	4	17.5	27	1	38	65	6	16.6	29	1	56	85	8	15.5
401–500	16	0	34	50	4	17.1	29	1	51	80	7	15.5	30	1	70	100	9	14.9
501–600	16	0	34	50	4	17.1	31	1	64	95	8	14.3	43	2	72	115	10	13.9
601–800	17	0	43	60	5	16.2	32	1	78	110	9	13.9	45	2	90	135	12	13.5
801–1000	30	1	45	75	6	15.0	45	2	75	120	10	13.3	60	3	110	170	14	12.4
1001–2000	31	1	59	90	7	14.5	50	2	100	150	12	12.7	75	4	160	235	19	11.5
2001–3000	32	1	68	100	8	14.0	50	2	130	180	14	12.0	95	5	185	280	22	11.0
3001–4000	34	1	81	115	9	13.5	65	3	135	200	15	11.3	95	5	255	350	27	10.5
4001–5000	35	1	95	130	10	13.0	70	3	155	225	17	11.0	130	7	260	390	29	10.0
5001–7000	50	2	90	140	11	12.5	70	3	185	255	19	10.7	160	9	355	515	38	9.5
7001–10,000	50	2	105	155	12	12.1	85	4	200	285	21	10.4	180	10	430	610	44	9.2
10,001–20,000	50	2	125	175	13	11.7	100	5	220	320	23	10.0	215	12	490	705	50	8.9
20,001–50,000	50	2	135	185	14	11.3	120	6	290	410	29	9.5	230	13	605	835	59	8.7
50,001–100,000	55	2	160	215	16	11.0	140	7	315	455	32	9.3	265	15	705	970	68	8.5

Trial 1: n_1 = first sample size; c_1 = acceptance number for first sample
"All" indicates that each piece in the lot is to be inspected
Trial 2: n_2 = second sample size; c_2 = acceptance number for first and second samples combined
p_t = lot tolerance per cent defective with a Consumer's Risk (P_C) of 0.10

Double Sampling Table for Average Outgoing Quality Limit (AOQL) = 7.0%

Lot Size	Process Average 0 to 0.14%						Process Average 0.15 to 1.40%						Process Average 1.41 to 2.80%					
	Trial 1		Trial 2			p_t %	Trial 1		Trial 2			p_t %	Trial 1		Trial 2			p_t %
	n_1	c_1	n_2	n_1+n_2	c_2		n_1	c_1	n_2	n_1+n_2	c_2		n_1	c_1	n_2	n_1+n_2	c_2	
1–5	All	0	–	–	–	–	All	0	–	–	–	–	All	0	–	–	–	–
6–25	5	0	–	–	–	35.5	5	0	–	–	–	35.5	5	0	–	–	–	35.5
26–50	7	0	4	11	1	38.5	7	0	4	11	1	38.5	7	0	4	11	1	38.5
51–100	7	0	5	12	1	36.5	7	0	5	12	1	36.5	9	0	9	18	2	31.5
101–200	8	0	5	13	1	36.0	9	0	10	19	2	30.5	10	0	15	25	3	27.5
201–300	8	0	6	14	1	35.5	9	0	11	20	2	30.0	10	0	17	27	3	27.0
301–400	9	0	11	20	2	30.0	9	0	11	20	2	30.0	11	0	17	28	3	26.0
401–500	9	0	11	20	2	30.0	9	0	11	20	2	30.0	11	0	17	28	3	26.0
501–600	9	0	12	21	2	29.5	9	0	12	21	2	29.5	12	0	25	37	4	24.0
601–800	9	0	12	21	2	29.5	9	0	12	21	2	29.5	12	0	25	37	4	24.0
801–1000	9	0	12	21	2	29.5	11	0	18	29	3	25.5	12	0	25	37	4	24.0
1001–2000	10	0	11	21	2	29.0	11	0	18	29	3	25.5	12	0	25	37	4	24.0
2001–3000	10	0	11	21	2	29.0	11	0	19	30	3	25.5	13	0	34	47	5	21.5
3001–4000	10	0	11	21	2	29.0	11	0	19	30	3	25.5	13	0	34	47	5	21.5
4001–5000	10	0	11	21	2	29.0	11	0	19	30	3	25.5	13	0	35	48	5	21.5
5001–7000	10	0	12	22	2	28.5	11	0	19	30	3	26.0	14	0	46	60	6	20.0
7001–10,000	10	0	12	22	2	28.5	12	0	26	38	4	23.5	25	1	50	75	8	18.5
10,001–20,000	10	0	12	22	2	28.5	12	0	27	39	4	23.0	25	1	55	80	8	18.0
20,001–50,000	10	0	12	22	2	28.5	13	0	36	49	5	21.0	25	1	60	85	9	17.8
50,001–100,000	10	0	12	22	2	28.5	13	0	36	49	5	21.0	26	1	69	95	10	17.5

Lot Size	Process Average 2.81 to 4.20%						Process Average 4.21 to 5.60%						Process Average 5.61 to 7.00%					
	Trial 1		Trial 2			p_t %	Trial 1		Trial 2			p_t %	Trial 1		Trial 2			p_t %
	n_1	c_1	n_2	n_1+n_2	c_2		n_1	c_1	n_2	n_1+n_2	c_2		n_1	c_1	n_2	n_1+n_2	c_2	
1–5	All	0	–	–	–	–	All	0	–	–	–	–	All	0	–	–	–	–
6–25	5	0	–	–	–	35.5	5	0	–	–	–	35.5	5	0	–	–	–	35.5
26–50	7	0	4	11	1	38.5	8	0	8	16	2	34.0	8	0	8	16	2	34.0
51–100	9	0	9	18	2	31.5	10	0	14	24	3	28.5	10	0	14	24	3	28.5
101–200	10	0	15	25	3	27.5	11	0	22	33	4	25.5	18	1	28	46	6	24.0
201–300	11	0	24	35	4	25.0	19	1	28	47	6	23.5	20	1	40	60	8	22.0
301–400	11	0	25	36	4	24.5	20	1	35	55	7	22.0	29	2	41	70	9	21.0
401–500	12	0	32	44	5	23.0	22	1	43	65	8	20.5	31	2	59	90	11	19.5
501–600	21	1	34	55	6	21.0	23	1	52	75	9	19.5	32	2	63	95	12	19.0
601–800	22	1	43	65	7	20.0	32	2	53	85	10	19.0	41	3	74	115	14	18.3
801–1000	23	1	42	65	7	19.8	33	2	62	95	11	18.4	43	3	82	125	15	17.5
1001–2000	23	1	47	70	8	19.2	34	2	81	115	13	17.8	55	4	115	170	20	16.5
2001–3000	24	1	56	80	9	18.7	44	3	91	135	15	17.0	65	5	135	200	23	16.0
3001–4000	24	1	66	90	10	18.2	47	3	103	150	16	16.0	75	6	175	250	28	15.0
4001–5000	26	1	74	100	11	17.7	50	3	135	185	19	15.0	100	8	225	325	35	14.0
5001–7000	37	2	78	115	12	16.5	65	4	155	220	22	14.0	115	9	265	380	40	13.5
7001–10,000	38	2	87	125	13	16.2	75	5	180	255	25	13.6	130	10	310	440	45	13.0
10,001–20,000	38	2	97	135	14	16.0	85	6	210	295	29	13.2	150	12	370	520	52	12.5
20,001–50,000	38	2	107	145	15	15.8	100	7	235	335	33	12.9	180	14	460	640	63	12.0
50,001–100,000	39	2	111	150	16	15.7	110	8	300	410	40	12.6	205	16	555	760	75	11.7

Trial 1: n_1 = first sample size; c_1 = acceptance number for first sample
"All" indicates that each piece in the lot is to be inspected
Trial 2: n_2 = second sample size; c_2 = acceptance number for first and second samples combined
p_t = lot tolerance per cent defective with a Consumer's Risk (P_C) of 0.10

10.0% AOQL

Double Sampling Table for Average Outgoing Quality Limit (AOQL) = 10.0%

Lot Size	Process Average 0 to 0.20% Trial 1 n_1	c_1	Trial 2 n_2	n_1+n_2	c_2	p_t %	Process Average 0.21 to 2.00% Trial 1 n_1	c_1	Trial 2 n_2	n_1+n_2	c_2	p_t %	Process Average 2.01 to 4.00% Trial 1 n_1	c_1	Trial 2 n_2	n_1+n_2	c_2	p_t %
1–3	All	0	—	—	—	—	All	0	—	—	—	—	All	0	—	—	—	—
4–15	3	0	—	—	—	50.0	3	0	—	—	—	50.0	3	0	—	—	—	50.0
16–50	5	0	3	8	1	53.5	5	0	3	8	1	53.5	5	0	3	8	1	53.5
51–100	5	0	3	8	1	55.0	6	0	8	14	2	43.0	6	0	8	14	2	43.0
101–200	5	0	4	9	1	52.0	7	0	7	14	2	42.0	7	0	12	19	3	38.0
201–300	7	0	7	14	2	42.5	7	0	7	14	2	42.5	7	0	13	20	3	37.0
301–400	7	0	7	14	2	42.5	7	0	7	14	2	42.5	8	0	17	25	4	35.0
401–500	7	0	8	15	2	40.0	7	0	8	15	2	40.0	8	0	18	26	4	34.0
501–600	7	0	8	15	2	40.0	8	0	13	21	3	35.0	8	0	18	26	4	34.0
601–800	7	0	8	15	2	40.5	8	0	13	21	3	35.0	8	0	18	26	4	34.5
801–1000	7	0	8	15	2	40.5	8	0	13	21	3	35.0	9	0	18	27	4	33.0
1001–2000	7	0	8	15	2	40.5	8	0	14	22	3	34.0	9	0	23	32	5	31.0
2001–3000	7	0	8	15	2	41.0	8	0	14	22	3	34.0	9	0	24	33	5	30.0
3001–4000	7	0	8	15	2	41.0	8	0	14	22	3	34.5	9	0	24	33	5	30.5
4001–5000	7	0	8	15	2	41.0	8	0	14	22	3	35.0	10	0	29	39	6	29.5
5001–7000	7	0	8	15	2	41.0	9	0	18	27	4	32.5	16	1	29	45	7	28.5
7001–10,000	7	0	8	15	2	41.0	9	0	18	27	4	32.5	17	1	38	55	8	26.0
10,001–20,000	7	0	8	15	2	41.0	9	0	18	27	4	32.5	17	1	38	55	8	26.0
20,001–50,000	7	0	8	15	2	41.0	9	0	18	27	4	32.5	18	1	42	60	9	25.5
50,001–100,000	8	0	14	22	3	33.5	9	0	25	34	5	30.0	18	1	52	70	10	24.5

Lot Size	Process Average 4.01 to 6.00% Trial 1 n_1	c_1	Trial 2 n_2	n_1+n_2	c_2	p_t %	Process Average 6.01 to 8.00% Trial 1 n_1	c_1	Trial 2 n_2	n_1+n_2	c_2	p_t %	Process Average 8.01 to 10.00% Trial 1 n_1	c_1	Trial 2 n_2	n_1+n_2	c_2	p_t %
1–3	All	0	—	—	—	—	All	0	—	—	—	—	All	0	—	—	—	—
4–15	3	0	—	—	—	50.0	3	0	—	—	—	50.0	3	0	—	—	—	50.0
16–50	6	0	6	12	2	48.0	6	0	6	12	2	48.0	6	0	6	12	2	48.0
51–100	7	0	11	18	3	38.5	7	0	11	18	3	38.5	7	0	16	23	4	36.5
101–200	8	0	16	24	4	35.5	13	1	20	33	6	33.5	14	1	24	38	7	32.0
201–300	8	0	17	25	4	35.0	14	1	26	40	7	31.5	19	2	29	48	9	31.0
301–400	8	0	22	30	5	34.0	15	1	30	45	8	31.0	21	2	44	65	12	29.0
401–500	15	1	23	38	6	30.5	16	1	39	55	9	28.5	22	2	53	75	13	27.0
501–600	16	1	28	44	7	28.5	22	2	38	60	10	27.5	28	3	52	80	14	26.5
601–800	16	1	28	44	7	29.0	22	2	43	65	11	27.0	29	3	56	85	15	26.0
801–1000	16	1	34	50	8	28.0	24	2	56	80	13	25.5	36	4	69	105	18	24.5
1001–2000	17	1	38	55	9	27.5	24	2	61	85	14	25.0	45	5	95	140	23	23.0
2001–3000	17	1	48	65	10	26.0	33	3	72	105	16	23.0	50	6	115	165	27	22.0
3001–4000	24	2	46	70	11	25.0	41	4	99	140	21	21.5	70	8	150	220	34	20.5
4001–5000	26	2	54	80	12	23.5	44	4	111	155	22	20.0	80	9	195	275	41	19.0
5001–7000	27	2	63	90	13	22.5	50	5	120	170	24	19.5	90	10	240	330	47	18.0
7001–10,000	27	2	68	95	14	22.0	60	6	145	205	28	18.5	110	12	265	375	53	17.5
10,001–20,000	28	2	77	105	15	22.0	70	7	165	235	32	18.0	125	14	320	445	62	17.0
20,001–50,000	28	2	87	115	17	21.5	80	8	205	285	39	17.5	140	16	355	495	69	16.8
50,001–100,000	36	3	99	135	20	21.0	85	8	245	330	44	17.0	150	17	390	540	77	16.6

Trial 1: n_1 = first sample size; c_1 = acceptance number for first sample
"All" indicates that each piece in the lot is to be inspected
Trial 2: n_2 = second sample size; c_2 = acceptance number for first and second samples combined
p_t = lot tolerance per cent defective with a Consumer's Risk (P_C) of 0.10

Index

A

Acceptable Quality Level (AQL), 3, 5
Acceptance criterion
    nature of, used in tables, 22
    for resubmitted lots, 52
Acceptance number, definition, 22
Acceptance Number, meaning of, 12, 22
Acceptance numbers, curves for determining
    AOQL single sampling (Fig. 2–9), 40
    LTPD double sampling (Fig. 2–7), 36
    LTPD single sampling (Fig. 1–2), 14
Acceptance number zones
    for AOQL single sampling plans, 39, 40
    for LTPD single sampling plans, 14
Accepted standards of good workmanship, 22
Accuracy of computed probabilities, 59–60
AOQL, numerical value of
    choosing, 6
    errors in, in tables, 42
    general considerations in choosing, 46
    practical examples of, 6
    review of, in shop applications, 52–53
    simple formula for, in single sampling, 37
AOQL sampling plans
    conditions best adapted to, 2, 30
    double sampling
        reasons for preference, 2, 5
        tables (Appendix 7), 205–218
    initial, 1, 17, 25
    mathematical relations, 37, 39–41
    practical need for, 25
    shop procedure in applying, 44–54
    single sampling tables (Appendix 6), 197–204
    usefulness of, 2
AOQL vs. LTPD, choice of, 5, 30–31
Application of tables, general field of, 2, 21–23
Approximate formulas, errors due to use of, 41–42
Approximations
    for binomial probabilities, 33, 60
    for hypergeometric probabilities, 33, 60
ASA Standards Z1.1 and Z1.2, 49
Assurance of conformance to requirements, use of tables
    to provide, 21

B

A.S.T.M., 43
Attributes, method of, meaning, 11, 22
"Average," used to signify "expected," 13
Average number inspected per lot
    mathematical formulas for, 34
    minimum, 13, 25, 26, 28
        example, LTPD single sampling, 15–16, 26
Average Outgoing Quality (AOQ), 37
    curves of (Fig. 2–4), 29
    expected value in practice, relative to AOQL, 32
Average Outgoing Quality Limit (AOQL) (see also AOQL)
    description and meaning of, 24
    initial concept of, 1, 17
Average quality protection
    concept of, 24
    double sampling AOQL tables (Appendix 7), 205–218
    general treatment of, 29–30
    mathematical relations for, 37, 39–41
    single sampling AOQL tables (Appendix 6), 197–204

B

Ballistic Research Laboratories, 55
Bartky, Walter, 39
Bell Telephone Laboratories, 1, 18, 32, 34, 39, 44, 60
Bender, Ruth A., 32
Binomial, 18–19, 33, 55, 59, 60
    formula for, 33
Binomial distribution, 13, 60
Binomial distribution of lot quality, 56
Breining, Elizabeth (Lockey), 60

C

Campbell, G. A., 19, 43
Campbell, George C., 39
Characteristic(s)
    how many to include, 3, 4–5
    treatment of a group of, 4, 30
    treatment of a single, 4–5, 30
Charts
    for finding AOQL single sampling plans, 40
    for finding LTPD double sampling plans, 36, 38
    (continued)